HAWKEYE COMMUNITY COLLEGE

3 7944 1006 2209 7

D1191649

WITHDRAWN

6
A
S
st
42

SIMPLIFIED DESIGN
OF STEEL STRUCTURES

Other titles in the
PARKER–AMBROSE SERIES OF SIMPLIFIED DESIGN GUIDES

624.1821
A496

3-3478

37.95

SIMPLIFIED DESIGN OF STEEL STRUCTURES

Seventh Edition

JAMES AMBROSE
Formerly Professor of Architecture
University of Southern California
Los Angeles, California

based on the work of

THE LATE HARRY PARKER
Formerly Professor of Architectural Construction
University of Pennsylvania

JOHN WILEY & SONS, INC.
New York • Chichester • Weinheim • Brisbane • Singapore • Toronto

042520

This text is printed on acid-free paper.

Copyright © 1997 by John Wiley & Sons, Inc.

All rights reserved. Published simultaneously in Canada.

Reproduction or translation of any part of this work beyond
that permitted by Section 107 or 108 of the 1976 United
States Copyright Act without the permission of the copyright
owner is unlawful. Requests for permission or further
information should be addressed to the Permissions Department,
John Wiley & Sons, Inc., 605 Third Avenue, New York, NY
10158-0012.

This publication is designed to provide accurate and
authoritative information in regard to the subject
matter covered. It is sold with the understanding that
the publisher is not engaged in rendering legal, accounting,
or other professional services. If legal advice or other
expert assistance is required, the services of a competent
professional person should be sought.

Library of Congress Cataloging-in-Publication Data:

Ambrose, James E.
   Simplified design of steel structures / James E. Ambrose ; based
on the work of the late Harry Parker. — 7th ed.
     p.   cm. — (Parker-Ambrose series of simplified design guides)
   Includes index.
   ISBN 0-471-16574-3 (cloth : alk. paper)
   1. Building, Iron and steel. 2. Steel, Structural. I. Parker,
Harry, 1887– Simplified design of steel structures. II. Title.
III. Series.
TA684.A343   1997
624.1′821—dc21   97-51

Printed in the United States of America

10 9 8 7 6 5 4 3 2 1

CONTENTS

PREFACE TO THE SEVENTH EDITION

This book is for persons interested in the design of steel structures for buildings. Because some readers may not be well versed in engineering analysis, applied mechanics, and fundamentals of structural behaviors, I have kept the mathematics, computational procedures, and general analytical complexity at a relatively undemanding level. This book is not essentially about performing computations, but about designing structures, which encompasses many things beyond the mathematical investigations of behaviors.

A significant portion of this work is based on texts originally prepared by Professor Harry Parker, who wanted to make his work accessible to persons with limited training in engineering analysis. Part of Professor Parker's preface to the first edition of this book is reprinted following this Preface. Since the first edition was published in 1945, Professor Parker and others (including me for the previous two editions) have periodically updated the book to reflect current practices and experiences in the steel products industry and the building design field in general. But keeping up with the times is a daunting challenge, because developing resources for steel production, refining processes for design and construction, and researching to support new practices are ongoing, enormous undertakings.

The material presented in this edition conforms with the latest edition of the widely accepted AISC Specification, published by the American Institute of Steel Construction. At present the AISC publishes two manuals, each with a different specification: one supports the traditional allowable stress design (ASD) method and the other supports the load and resistance factor design (LRFD) method. (Both manuals contain extensive data to support design work.) Most of the design work in this book follows the forms developed for the simpler ASD method because I want to explain basic design issues, not investigative methods, and I want to do so as simply as possible. However, because the LRFD method is steadily becoming the method of choice for major design work, I present a basic explanation of it in Chapter 5. In addition, I illustrate many design processes in the LRFD method after I make a basic presentation in the ASD method.

What we use for building construction today is a complex array of old and new materials combined with various accessory elements. Compared to wood, masonry, and concrete, steel is a relatively new material used for building. However, we know a great deal about the use of steel—it has been around for 150 years. And current construction processes use all our experience and knowledge, as well as available technologies.

In this book I focus on the working context of the structural designer, specifically the designer of building structures. Many practical matters and specific needs tend to limit and focus the work, but what endures is a body of material that is truly required to define a design solution adequately for builders.

Because construction is highly repetitive in nature, much of the design "work" refers to predesigned elements and systems—possibly accessible from published tables, graphs, computer-aided design (CAD) programs, manufacturers catalogs, or designers' personal files. I show how to use some of these references through case studies.

This book is intended mostly for inexperienced designers, so it contains many carefully developed explanations of basic concepts and relationships. Readers with different backgrounds and specific interests may shop among these presentations to satisfy their particular needs. To further enhance the book's use as a self-study manual, I provide study aids at the back of the book.

I am grateful to several organizations for their permission to use materials from their publications: the International Conference of Building Officials (ICBO) for permission to use materials from the *Uniform Building Code*, the American Institute of Steel Construction (AISC) for permission to use materials from their manuals, and to the Steel Joist Institute (SJI) and the Steel Deck Institute (SDI) for permission to use some materials presented from their publications.

I am grateful to John Wiley & Sons for its continued interest in and publication of its series of practical design guides. I am especially grateful to Peggy Burns, Amanda Miller, and Mary Masi of the Professional, Reference, and Trade Group, and to Jennifer Mazurkie of the Wiley production division.

As always, I must also acknowledge my indebtedness to my professional partner, my wife, Peggy, without whose direct assistance and continued support this work would not continue.

JAMES AMBROSE

July 1997

PREFACE TO THE
FIRST EDITION

(The following is an excerpt from Professor Parker's preface to the first edition.)

Simplified Design of Structural Steel is the fourth of a series of elementary books dealing with the design of structural members used in the construction of buildings. The present volume treats of the design of the most common structural steel members that occur in building construction. The solution of many structural problems is difficult and involved but it is surprising, on investigation, how readily many of the seemingly difficult problems may be solved. The author has endeavored to show how the application of the basic principles of mechanics simplifies the problems and leads directly to a solution. Using tables and formulas blindly is a dangerous procedure; they can only be used with safety when there is a clear understanding of the underlying principles upon which the tables or formulas are based. This book deals principally in the practical application of engineering principles and formulas in the design of structural members.

In preparing material for this book the author has assumed that the reader is unfamiliar with the subject. Consequently the discussions advance by easy stages, beginning with problems relating to simple direct stresses and continuing to the more involved examples. Most of the fundamental principles of mechanics are reviewed and, in general, the only preparation needed is a knowledge of arithmetic and high school algebra.

In addition to discussions and explanations of design procedure, it has been found that the solution of practical examples adds greatly to the value of a book of this character. Consequently, a great portion of the text consists of the solution of illustrative examples. The examples are followed by problems to be solved by the student.

The author proposes no new methods of design nor short cuts of questionable value. Instead, he has endeavored to present concise and clear explanations of the present-day design methods with the hope that the reader may obtain a foundation of sound principles of structural engineering.

HARRY PARKER

High Hollow, Southampton, Pa.
March 1945

SIMPLIFIED DESIGN
OF STEEL STRUCTURES

1

INTRODUCTION

Designers who plan to use steel structures in buildings must consider a broad range of variables, including steel's properties and the common forms of industrialized steel products. Designers also must be aware of routine usages for typical building construction, general building design concerns, regulatory codes, and oft-used design methods. I briefly discuss some of these concerns in Chapters 1 and 2.

1.1 USE OF STEEL FOR BUILDING STRUCTURES

In general this book covers the common uses of steel for ordinary building structures. But since steel is used in some form in wood, concrete, and masonry structures—for example, wood frames require nails, screws, and bolts, as well as various devices, while modern concrete and masonry construction typically requires steel reinforcement, as well as various steel anchorage and attachment devices—I limit my discussions to

- Situations in which steel is the major material for the primary structure
- The use of steel spanning systems in combination with vertical bearing structures of concrete or masonry
- The use of composite spanning elements of steel and concrete or steel and wood

Many designers who use steel erect a primary frame consisting of structural steel; the term *structural steel* applies to structures in which the major elements are produced by *hot-rolling*, resulting in linear elements of constant cross section. In the United States rolled sections ordinarily are formed into shapes that conform with standards established by the American Institute of Steel Construction (AISC), as documented in its publication *Manual of Steel Construction* (Ref. 3), commonly known as the AISC Manual.

Other steel structures use elements a notch smaller and lighter than rolled products, commonly produced by *cold-forming* (stamping, folding, cold-rolling, etc.) thin steel sheets. Such products, which include roof and floor decking and some wall paneling, are often described as *formed sheet steel*. Light framed elements, such as open web joists, also may use cold-formed elements.

A third use of steel is *miscellaneous metals*, which includes elements that are not part of the primary structural system but serve some secondary structural function, such as framing for curtain walls, suspended ceilings, door frames, and so on.

Note: Light steel structures sometimes are made from very light rolled shapes, and miscellaneous metals sometimes are made from the smallest rolled shapes or cold-formed sheet steel.

In much of this book I deal with structural steel elements and systems. However, I also detail the major uses of cold-formed sheet steel.

1.2 METHODS OF INVESTIGATION AND DESIGN

Nowadays designers may use either of two fundamentally different methods for structural investigation and design. The first, traditionally used by designers and researchers, is the *working stress method*, or the *allowable stress method*. At present this method is called the *allowable stress design* (ASD) method. The second method, now gaining in favor, is the *ultimate strength method*, or the *strength method*. At present this method is called the *load and resistance factor design* (LRFD) method.

In general ASD techniques and operational procedures are simpler to use—and to demonstrate. They are based largely on direct use of

- Classical analytical formulas for stress and strain
- Actual working loads (called *service loads*) assumed for the structure

In fact, many people explain the LRFD's analytical methods by comparing them to ASD methods.

Because most of the work in this book involves simple and ordinary structures, I use the ASD method to explain basic problems. However, I also explain the basic applications of the LRFD method as often as possible to give the reader an opportunity to see the differences between the methods.

The AISC presently supports both methods, publishing separate references (Ref. 3 and 4). Samples of data from the AISC Manuals are presented throughout this book for use in example computations and exercise problems. The data relating to the work in this book are largely unchanged from editions of the manuals prior to the most recent, so, as noted, some data is adapted from previous editions. For any actual design work, the reader should use as a reference the current edition of the AISC Manuals where the data base is considerably more extensive and may contain new materials.

In sum, for practical purposes, the ASD method is simpler and better for persons learning basic issues; use it as a stepping stone to the more complex LRFD method. Most designers expect the LRFD method to prevail in time if modified to be more intelligible and user friendly.

Note: I discuss the bases for the methods and I describe how to choose between them in Chapter 5.

1.3   REFERENCE SOURCES FOR DESIGN

The information presented in this book is general in nature, specific to the interests of building designers, and represents a small, distilled essence of a number of general publications. The book's principal references are listed in the Bibliography.

Designers can find information about steel structures from a number of sources, ranging from relatively unbiased textbooks and research reports to biased promotional materials from the producers and suppliers of construction materials, equipment, and services. (One cannot expect people in the business of selling steel and fabricated steel products to provide unbiased information about steel's drawbacks.) Some publications are somewhat more essential and general in nature; most are highly detailed and narrowly directed.

The sources used most by designers of steel structures are the AISC manuals (Ref. 3 and 4) I described in the preceding section. These manuals include the latest design specification, indispensable data regarding currently available steel products for structures, and many shortcut design aids (permitting rapid design of commonly used structural elements).

Throughout this book I provide samples of materials found in the AISC manuals to help explain how to use the manuals for investigation and design of structures. Readers can perform all the exercise problems using the materials provided, but I advise you to obtain access to the current manuals so that you become familiar with the full scope of materials provided there.

Note: The AISC, which is the principal industry-sponsored organization dealing with steel structures, also publishes many other references for information about steel structures. Other industry and professional organizations that provide materials for designers are

> *The American Society for Testing and Materials (ASTM).* This organization provides standards for all sorts of materials, including many structural products. Just about every steel structural product has some ASTM specification.
>
> *The American Society of Civil Engineers (ASCE).* An ASCE division is the major organization of structural engineers and

sponsors many publications on structural design, including the current major guide for design loads: *Minimum Design Loads for Buildings and Other Structures* (Ref. 2).

The Steel Deck Institute (SDI). This group provides information for design of formed sheet steel products, widely used for roof and floor decks and wall panel units.

The Steel Joist Institute (SJI). This group provides information regarding light fabricated trusses (called *open-web joists*) and other forms of steel spanning members.

In sum, several notable industry-wide organizations relate somehow to steel structures. While these organizations are largely industry-supported, they do represent major sources of design codes and standards, as well as product information. They also sponsor much of the research on which design procedures are based.

In fact, an industry or trade organization exists for almost every type of product used for construction. Any of them may be the source of useful information. The ones I refer to here are just a sampling.

Anyone intending to study this subject beyond the scope of this book should obtain a basic text, such as those used in civil engineering schools. *Steel Buildings: Analysis and Design* (Ref. 5) and *Structural Steel Design: LRFD Method* are two such publications.

1.4 UNITS OF MEASUREMENT

Previous editions of this book used U.S. units (feet, inches, pounds, etc.). In this edition I again use U.S. units but list equivalent metric units in brackets [thus]. Although the U.S. building industry is slowly changing to metric units, my decision to use U.S. units is pragmatic: most of the references used for this book still prefer U.S. units.

Table 1.1 lists the standard units of measurement in the U.S. system with the abbreviations used in this work and a description of the units' common use in structural design work. In similar form, Table 1.2 gives the corresponding units in the metric system. Table 1.3 lists the conversion factors needed to shift from one system to the other.

TABLE 1.1 Units of Measurement: U.S. System

Unit	Abbreviation	Use in Building Design
Length		
Foot	ft	Large dimensions, building plans, beam spans
Inch	in.	Small dimensions, size of member cross sections
Area		
Square feet	ft^2	Large areas
Square inches	$in.^2$	Small areas, properties of cross sections
Volume		
Cubic yards	yd^3	Large volumes of soil or concrete (commonly called *yards*)
Cubic feet	ft^3	Quantities of material
Cubic inches	$in.^3$	Small volumes
Force, Mass		
Pound	lb	Specific weight, force, load
Kip	kip, k*	1000 pounds
Ton	ton	2000 pounds
Pounds per foot	lb/ft, plf	Linear load (as on a beam)
Kips per foot	kips/ft, klf	Linear load (as on a beam)
Pounds per square foot	lb/ft^2, psf	Distributed load on a surface, pressure
Kips per square foot	$kips/ft^2$, ksf	Distributed load on a surface, pressure
Pounds per cubic foot	lb/ft^3	Relative density, weight
Moment		
Foot-pounds	ft-lb	Rotational or bending moment
Inch-pounds	in.-lb	Rotational or bending moment
Kip-feet	kip-ft	Rotational or bending moment
Kip-inches	kip-in.	Rotational or bending moment
Stress		
Pounds per square foot	lb/ft^2, psf	Soil pressure
Pounds per square inch	$lb/in.^2$, psi	Stresses in structures
Kips per square foot	$kips/ft^2$, ksf	Soil pressure
Kips per square inch	$kips/in.^2$, ksi	Stresses in structures
Temperature		
Degree Fahrenheit	°F	Temperature

*k should be used only in abbreviations—ksi, klf, ksf—not as a stand-alone symbol.

Using a conversion factor produces a *hard conversion*—that is, a reasonably precise conversion. In this book however, many of the unit conversions presented are *soft conversions*—that is, I rounded off converted values to produce approximate equivalent values of slightly more relevant numerical significance to the unit system.

TABLE 1.2 Units of Measurement: Metric System

Unit	Abbreviation	Use in Building Design
Length		
Meter	m	Large dimensions, building plans, beam spans
Millimeter	mm	Small dimensions, size of member cross sections
Area		
Square meters	m^2	Large areas
Square millimeters	mm^2	Small areas, properties of cross sections
Volume		
Cubic meters	m^3	Large volumes
Cubic millimeters	mm^3	Small volumes
Mass		
Kilogram	kg	Mass of materials (equivalent to weight in U.S. units)
Kilograms per cubic meter	kg/m^3	Density (unit weight)
Force, Load		
Newton	N	Force or load
Kilonewton	kN	1,000 Newtons
Stress		
Pascal	Pa	Stress or pressure (1 pascal = $1 N/m^2$)
Kilopascal	kPa	1,000 pascals
Megapascal	MPa	1,000,000 pascals
Gigapascal	GPa	1,000,000,000 pascals
Temperature		
Degree Celsius	°C	Temperature

TABLE 1.3   Factors for Converting Between Unit Systems

To Convert from U.S. Units to Metric Units, Multiply by	U.S. Unit	Metric Unit	To Convert from Metric Units to U.S. Units, Multiply by
25.4	in.	mm	0.03937
0.3048	ft	m	3.281
645.2	in.^{2}	mm^2	1.550×10^{-3}
16.39×10^3	in.^{3}	mm^3	61.02×10^{-6}
416.2×10^3	in.^{4}	mm^4	2.403×10^{-6}
0.09290	ft^2	m^2	10.76
0.02832	ft^3	m^3	35.31
0.4536	lb (mass)	kg	2.205
4.448	lb (force)	N	0.2248
4.448	kip (force)	kN	0.2248
1.356	ft-lb (moment)	N-m	0.7376
1.356	kip-ft (moment)	kN-m	0.7376
16.0185	lb/ft^3 (density)	kg/m^3	0.06243
14.59	lb/ft (load)	N/m	0.06853
14.59	kips/ft (load)	kN/m	0.06853
6.895	psi (stress)	kPa	0.1450
6.895	ksi (stress)	MPa	0.1450
0.04788	psf (load or pressure)	kPa	20.93
47.88	ksf (load or pressure)	kPa	0.02093
$0.566 \times (^\circ F - 32)$	$^\circ F$	$^\circ C$	$(1.8 \times {^\circ C}) + 32$

For example, a wood 2 × 4 (actually 1.5 × 3.5 inches in the U.S. system) is precisely 38.1 × 88.9 mm in the metric system. However, the metric equivalent 2 × 4 is more likely to be 40 × 90 mm—close enough for most purposes in construction work.

1.5 COMPUTATIONAL ACCURACY

Structures for buildings are seldom produced precisely. Although some parts of the construction—such as window frames and elevator rails—must be dimensionally precise, the basic structural framework requires a very limited dimensional precision. Tack on a lack of precision in predicting loads and the significance of highly precise structural computations becomes moot. I am not justifying sloppy mathematical work, sloppy construction, or the use of vague theories when investigating behaviors. Nevertheless, I am not overly concerned with any numbers beyond the second digit.

While most professional designers these days use computers, most of the work illustrated here requires only a hand calculator (an eight-digit, scientific type is adequate). I sometimes round off even these primitive computations with no apologies.

1.6 SYMBOLS

Table 1.4 gives frequently used short shorthand symbols.

TABLE 1.4   Shorthand Symbols in Common Use

Symbol	Reading
$>$	is greater than
$<$	is less than
$\geqslant$	equal to or greater than
$\leqslant$	equal to or less than
$6'$	six feet
$6''$	six inches
Σ	sum of
Δ	change in

1.7 NOMENCLATURE

Notation used in this book complies generally with that used in the steel industry and the latest editions of standard specifications. The following list includes all the notation used in this book and is compiled and adapted from more extensive lists found in the references.

A = area, general

A_g = gross area of section, defined by the outer dimensions

A_n = net area

C = compressive force

E	=	modulus of elasticity of steel
F_a	=	allowable compressive stress due to axial load only
F_b	=	allowable compressive stress due to bending
F_u	=	specified minimum ultimate tensile strength of steel
F_y	=	specified minimum yield strength of steel
I	=	moment of inertia
K	=	factor for modifying unbraced length of column, based on support conditions
L	=	(1) length (usually of a span); (2) unbraced height of column
M	=	bending moment
M_R	=	resisting moment capacity of a section
P	=	concentrated load
S	=	section modulus
T	=	tension force
W	=	(1) total gravity load; (2) weight, or dead load, of an object; (3) total wind load force; (4) total of a uniformly distributed load or pressure due to gravity
b	=	width (general)
b_f	=	width of flange of W shape
c	=	in bending, distance from extreme fiber stress to the neutral axis
d	=	overall beam depth, out to out of flanges for a W shape
e	=	eccentricity of a nonaxial load, from point of application of the load to the centroid of the section
f_a	=	computed stress due to axial load
f_b	=	computed bending stress
f'_c	=	specified compressive strength of concrete
f'_m	=	specified compressive strength of masonry
f_p	=	computed bearing stress
f_t	=	computed stress in tension
f_v	=	computed shear stress
r	=	radius of gyration
s	=	spacing of objects, center to center
t	=	thickness, general

t_f =   thickness of flange of W shape

t_w =   thickness of web of W shape

w =   unit of a distributed load on a beam (e.g., lb/ft)

Δ =   deflection, usually maximum vertical deflection of a beam

Φ =   strength reduction factor in strength design

2

WHAT IS STEEL?

Hundreds of different steels exist, for steel's strength, hardness, and corrosion resistance (as well as other properties) can vary considerably, depending on production processes. In addition, working and forming processes—such as rolling, drawing, machining, and forging—also may alter some properties. In fact, only a few properties—including density (unit weight), stiffness (modulus of elasticity), thermal expansion, and fire resistance—tend to remain constant for all steels.

Although in reality a few standard steel products make up most of the elements of building structures, designers must consider steel's many inconstant properties, including

- Hardness, which affects how easily cutting, drilling, planing, and other working is done
- Weldability
- Rust resistance, which, though normally low, can be enhanced if you add various materials to the steel, producing special steels, such as stainless steel and the so-called *rusting steels*, which rust very slowly.

2.1 UNIQUE STRUCTURAL NATURE

Steel's basic structural properties, including strength, stiffness, ductility, and brittleness, can be interpreted from laboratory load tests; Figure 2.1 displays the characteristic curves obtained by plotting stress and strain values from such tests. Curve 1, which represents ordinary structural steel, illustrates an important property of many structural steels: plastic deformation (yield). For such steels, two stress values are significant: the *yield limit* and the *ultimate limit*.

Generally, the higher the yield limit, the less the degree of ductility. Ductility is the ratio of the plastic deformation between first yield and strain hardening to the elastic deformation at the point of yield. Curve 2 indicates the typical effect as the yield strength is raised: eventually, the significance of the yield phenomenon becomes virtually negligible when the yield strength approaches as much as three times the yield of ordinary steel.

Because steel is expensive, steel structures generally consist of elements with relatively thin parts. Some of the strongest steels are produced only in thin sheets or drawn wires. Bridge strand, for example, is made from wire; at 300,000 psi yield is almost nonexistent and the wires are as brittle as glass rods.

In such situations the ultimate limiting strength is determined by *buckling*, not the material's stress limits. Since buckling is a function of stiffness (modulus of elasticity), which is the same for all steels, designers cannot effectively use superstrong steels in many situations. In fact, the grades of steel most commonly used are ones that have the optimal effective strength for most typical tasks.

Note: Building designers rarely get to choose basic materials; the proper steel often is predetermined as part of a product design— many structural elements are produced as a product line.

Steel that meets the requirements of the American Society for Testing and Materials (ASTM) Specification A36 is the grade of structural steel most commonly used to produce rolled steel elements for building construction. Such steel, which must have an ultimate tensile strength of 58 to 80 ksi and a minimum yield point of 36 ksi, is used for bolted, riveted, or welded fabrication. This steel, which I refer to as A36 steel, is used for much of the work I describe in this book.

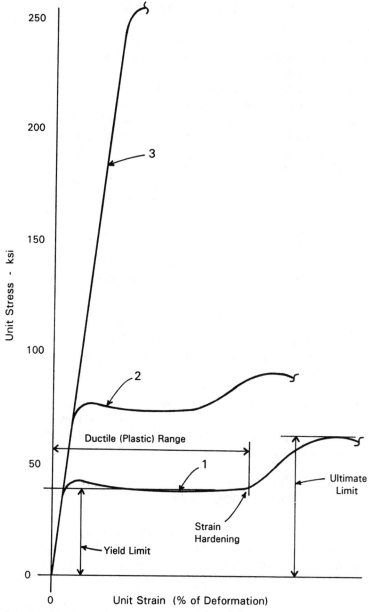

FIGURE 2.1 Stress/strain responses: 1) ordinary structural steel 2) high-strength steel for rolled shapes 3) super-strength steel (usually in wire form).

Prior to 1963 a steel designated ASTM A7 was the basic structural steel; with a yield point of 33 ksi, it was used primarily for riveted fabrication. With the increasing demand for bolted and welded construction, however, A36 steel became the material of choice for most structural products.

For structural steel the AISC Specification expresses the allowable unit stresses used for ASD work as some percent of the yield stress (F_y) or the ultimate stress (F_u). I describe allowable unit stresses used for design as appropriate where I present specific design problems. For more complete descriptions, check the AISC Specification in the AISC Manual (Ref. 3). In many cases there are a number of qualifying conditions for use of allowable stresses; I discuss some of these conditions elsewhere in the book.

LRFD work uses basic limiting stresses (yield and ultimate), not allowable stresses. You can modify stress conditions for specific usage situations (tension, bending, shear, etc.) by using different reduction factors called *resistance factors* (see Chapter 5).

Steel used for purposes other than the production of rolled products—for example, steel connectors, wire, cast and forged elements, and very high strength steels produced in sheet, bar, and rod form for fabricated products—generally conforms to standards developed for the specific product. (I discuss the properties and design stresses for some of these product applications in other sections of the book.) Such standards typically conform to those established by industry-wide organizations, such as the Steel Joist Institute (SJI) and the Steel Deck Institute (SDI). In some cases, larger fabricated products use ordinary rolled products produced from A36 steel (or other grades of steel from which hot-rolled products can be obtained).

2.2 INDUSTRIAL PROCESSES

Steel is essentially formless; initially it appears as a molten material or a softened lump. Steel products derive their basic forms during the production process. Standard raw stock elements and production processes are the following:

Rolled Shapes. Formed by squeezing heat-softened steel repeatedly through a set of rollers. The rollers shape the steel into a linear element with a constant cross section. Simple shapes include round rods and flat bars, strips, plates, and sheets; complex shapes include I, H, T, L, U, C, and Z; and special shapes include rails or sheet piling.

Wire. Formed by pulling (drawing) steel through a small opening.

Extrusion. Similar to drawing but the sections produced are not simple round shapes. This process is rarely used to develop building construction elements.

Casting. Pouring molten steel into a form (mold) produces three-dimensional objects. This process is rarely used to develop building construction elements.

Forging. Pounding softened steel into a mold until it takes the shape of the mold. Forging is preferred to casting because of its effects on the finished material.

Raw stock steel elements produced by basic forming processes may be reworked by various means:

Cutting. Shearing, sawing, punching, or flame cutting to trim and shape.

Machining. Drilling, planing, grinding, routing, or turning on a lathe.

Bending. The more ductile, the easier to bend.

Stamping. Punching sheet steel into a mold to produce a three-dimensional shape, such as a hemisphere.

Rerolling. Reworking a linear element into a curved form (arched) or forming a sheet or flat strip into a formed cross section (corrugated, etc.).

Raw stock or reformed elements may be assembled into an object of multiple parts (such as a manufactured truss or a prefabricated wall panel) by various means:

Fitting. Screwing together threaded parts. Also refers to interlocking techniques, such as the tongue-and-groove joint or the bayonet twist lock.

Friction. Clamping, wedging, or squeezing with high-tensile bolts to resist the sliding of parts in surface contact.

Pinning. Placing a pin-type device (bolt, rivet, or actual pin) through matching holes in overlapping flat elements to prevent the slipping of parts at the contact face.

Nailing, Screwing. Attaching thin elements—most will have some preformed holes—by nails or screws.

Welding. Melting elements together at the contact point to produce a bonded connection. Includes gas and electric arc welding.

Adhesive Bonding. Chemical bonding to fuse the materials of connected parts.

2.3 ROLLED STRUCTURAL SHAPES

Hot-rolled steel products used as beams, columns, and other structural members have distinct *shapes.* How they are used is related to the profiles of their cross sections.

American standard I-beams (see Figure 2.2*a*), the first beam sections rolled in the United States, have a depth of 3 to 24 in. W shapes (see Figure 2.2*b*), originally called *wide-flange shapes,* have a depth of 4 to 44 in. The W shape is a modification of the I cross section; W shapes have parallel flange surfaces, while standard I-beams have tapered inside flange surfaces (the inside faces slope $16\frac{2}{3}\%$, or 1 in 6).

Other structural steel shapes commonly used in building construction are channels, angles, tees, plates, and bars. The tables in Appendix A list the dimensions, weights, and various properties of some of these shapes. For complete tables of structural shapes, refer to the AISC Manual (Ref. 3).

Standard I-Beams (S Shapes)

American standard I-beams are identified by the letter S, followed by the depth in inches and the weight in pounds per linear foot. The designation S 12 × 35 thus indicates a standard shape 12 in. deep weighing 35 lb per linear foot. Standard I-beams in a given depth group have uniform depths because rolls are spread in one direction only to make shapes of greater cross-sectional area. Although the depth remains constant, flange width and web thickness increase.

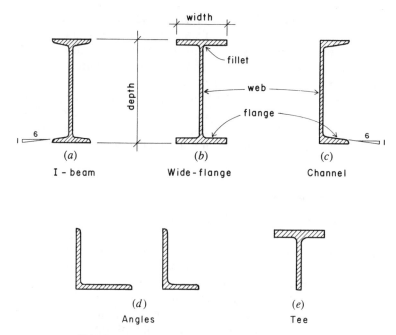

FIGURE 2.2 Shapes of typical hot-rolled products.

In general S shapes are not as structurally efficient as W shapes and so are not as widely used. S shapes are used when a situation calls for constant depth, narrow flanges, and thicker webs.

W Shapes

W shapes are identified by the letter W, followed by the *nominal* depth in inches and the weight in pounds per linear foot. Thus the designation W 12 × 26 indicates a wide-flange shape nominally 12 in. deep weighing 26 lb per linear foot.

A W shape's actual depth varies within nominal depth groupings. A W 12 × 26 has an actual depth of 12.22 in., for example, whereas a W 12 × 45 is actually 12.06 in. deep (see Table A.1 in Appendix A). The actual depths are different than the nominal depths because rolls are spread vertically and horizontally to make shapes of greater cross-sectional areas. In general W shapes have greater flange widths and relatively thinner webs than standard I-beams. More material in the flanges makes wide-flange shapes more efficient structurally than standard I-beams. In fact, many wide-flange shapes are rolled so that flange widths approximately equal their depths. The resulting cross sections (H configurations) are much more suitable for use as columns than the I profiles.

I recommend that the reader compare shapes in their respective nominal depth groups to become familiar with the variety of geometrical relationships. For example, note that a wide variety of weights is available within each nominal depth group.

Standard Channels

American standard channels (see Figure 2.2c) are identified by the letter C. The designation C 10 × 20 indicates a standard channel 10 in. deep weighing 20 lb per linear foot. Like the standard I-beams, the depth of a particular group remains constant because rolls are spread in one direction to increase the cross-sectional area. Because channels tend to buckle when used independently as beams or columns, they require lateral support or bracing. Although channels are generally used as elements of built-up sections, such as columns and lintels, the absence of a flange on one side makes channels particularly suitable for framing around floor openings.

Angles

Structural angles have L-shaped sections. Both legs of an angle have the same thickness. Table A.2 in Appendix A gives dimensions, weights, and other properties of equal and unequal leg angles.

Angles are identified by the letter L, followed by the dimensions of the legs and their thickness. Thus the designation L 4 × 4 × $\frac{1}{2}$ indicates an equal leg angle with 4-in. legs, $\frac{1}{2}$ in. thick. From Table A.2, this section weighs 12.8 lb per linear foot and has a cross-sectional area of 3.75 in.^{2}. Similarly, the designation L 5 × $3\frac{1}{2}$ × $\frac{1}{2}$ indicates an unequal leg angle with a 5-in. and a $3\frac{1}{2}$-in. leg, both $\frac{1}{2}$ in. thick. From Table A.2, this angle weighs 13.6 lb per linear foot and has an area of 4 in.^{2}.

To change the weight and area of an angle of a given leg length, increase the thickness of each leg the same amount. For example, if the leg thickness of an L $5 \times 3\frac{1}{2} \times \frac{1}{2}$ is increased to $\frac{5}{8}$ in., Table A.2 shows that the resulting angle (L $5 \times 3\frac{1}{2} \times \frac{5}{8}$) weighs 16.8 lb per linear foot has and an area of 4.92 in.^{2}. *Note:* This method of spreading the rolls changes the leg lengths slightly.

Before the advent of the heavier W shapes, angles were used as elements of built-up sections, such as plate girders and heavy columns. Now single angles often are used as lintels; double angles often are used as members of light steel trusses. Short lengths of angles commonly serve as connecting members for beams and columns.

Structural Tees

To make a structural tee, split the web of a W shape (Fig. 2.2*e*) or a standard I-beam (S shape). The cut, normally made along the web's center, produces tees with a stem depth equal to half the depth of the original section. Structural tees cut from W shapes are identified by the letters WT; those cut from standard S shapes, by ST. The designation WT 6×53 indicates a structural tee 6 in. deep weighing 53 lb per linear foot; such a shape is produced by splitting a W 12×106. Similarly, ST 9×35 indicates a structural tee 9 in. deep weighing 35 lb per linear foot and cut from an S 18×70. Structural tees are used as chord members of welded steel trusses and as flanges in certain plate girders.

Bars and Plates

Flat structural steel elements are generally classified as follows:

Bars. Width $\leqslant 6$ in., thickness $\geqslant .203$ in.
Plates. Width > 8 in., thickness $\geqslant .230$ in.; or width > 48 in., thickness $\geqslant .180$ in.
Sheets. Thickness $< .180$ in.

Bars and plates come in many different sizes. Bars are available in many widths and virtually all thicknesses and lengths. The usual practice is to specify bars in increments of $\frac{1}{4}$ in. for widths and $\frac{1}{8}$ in. for thickness. The usual practice is to specify plates as follows:

Widths. Increments of even inches.
Thickness. $\frac{1}{32}$-in. increments up to $\frac{1}{2}$ in.
  $\frac{1}{16}$-in. increments between $\frac{1}{2}$ in. and 2 in.
  $\frac{1}{8}$-in. increments between 2 in. and 6 in.
  $\frac{1}{4}$-in. increments greater than 6 in.

  The standard dimensional sequence when describing steel plate is

$$\text{thickness} \times \text{width} \times \text{length}$$

All dimensions are given in inches, fractions of an inch, or decimals of an inch.

  I discuss the design of column-base plates and beam-bearing plates in Sections 7.13 and 6.10, respectively.

Other Designations for Structural Steel Elements

As I noted earlier in this section, American standard beam shapes are identified by the letter S, while wide-flange shapes are identified by the letter W. A third designation, M, covers miscellaneous shapes that cannot be classified as W or S; such shapes have various slopes on their inner flange surfaces.

  Similarly, the letters MC identify rolled channels that cannot be classified as C shapes.

  Table 2.1 lists the standard designations used for rolled shapes, formed rectangular tubing, and round pipe.

2.4 COLD-FORMED STEEL PRODUCTS

To produce elements from sheet steel, you ordinarily do not need to heat the steel; thus these elements are known as *cold-formed*. Because cold-formed products typically are formed from very thin sheets, they also are called *light-gage* steel products. Figure 2.3 shows the cross sections of some cold-formed products.

TABLE 2.1 Standard Designations for Structural Steel Elements

Element	Designation
American standard I-beams (S shapes)	S 12 × 35
Wide-flange (W) shapes	W 12 × 27
Miscellaneous shapes	M 8 × 18.5
American standard channels	C 10 × 20
Miscellaneous channels	MC 12 × 45
Angles	L 5 × 3 × $\frac{1}{2}$
Structural tees (cut from W shapes)	WT 6 × 53
Plate	PL $\frac{1}{2}$ × 12 × 1'–4"
Structural tubing	TS 4 × 4 × 0.375
Pipe (standard weight)	Pipe 4 std.

Large corrugated or fluted sheets often are used for paneling and for structural decks (roofs and floors). I discuss using these elements for floor decking in Chapter 6. A number of manufacturers make these products; for information regarding their structural properties, contact the manufacturer directly. For general information on structural decks, check the *Steel Deck Institute Design Manual for Composite Decks, Form Decks, and Roof Decks* (Ref. 8).

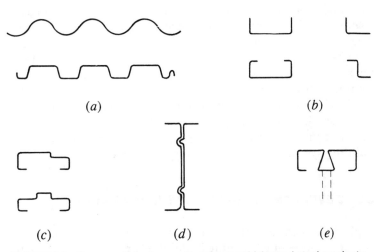

(a) *(b)*

(c) *(d)* *(e)*

FIGURE 2.3 Cross section shapes of common cold-formed steel products.

Cold-formed shapes range from the simple L, C, U, and so on to the special forms produced for various construction systems. Some building structures are almost entirely composed of cold-formed products. Several manufacturers produce systems of these components—voilà, predesigned, packaged building structures. The design of cold-formed elements is described in the *Cold-Formed Steel Design Manual*, published by the American Iron and Steel Institute.

2.5 FABRICATED STRUCTURAL COMPONENTS

A number of special steel products used as structural members combine hot-rolled and cold-formed elements.

Open-web steel joists consist of prefabricated, light steel trusses. For short spans and light loads, the web commonly consists of a single, continuous bent steel rod and the chords of steel rods or cold-formed elements (see Figure 2.4a). For larger spans or heavier loads, the forms resemble those used for ordinary light steel trusses—single angles, double angles, and structural tees constitute the truss members. I discuss open-web joists for floor framing in Section 6.11.

The fabricated joist shown in Figure 2.4b is formed by cutting the web of standard rolled shapes in a zigzag fashion (see Figures 2.4c and d). The resulting product boasts a reduced weight-to-depth ratio, compared to the rolled shapes.

Other fabricated steel products range from whole building systems to elements for windows, doors, curtain wall systems, and the framing for interior partition walls. Some components and systems are developed under controls of industry-wide standards as industry guides, but many are proprietary items produced by a single manufacturer. For examples of such structural products, see the building system design examples in Chapter 12.

2.6 DEVELOPMENT OF STRUCTURAL SYSTEMS

Typically structural systems comprising entire roof, floor, or wall constructions—or even entire buildings—are assembled from individual elements. For example, a typical floor combines rolled

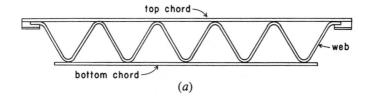

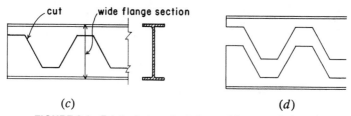

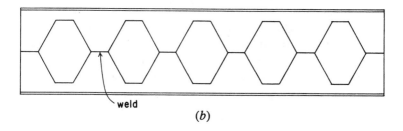

FIGURE 2.4 Fabricated products formed from steel elements.

steel beams and a formed sheet steel deck. Designers may perform structural investigations, but often the choice of elements is largely a practical outgrowth of the construction form.

Most buildings incorporate more than one material. Various combinations are possible, such as a wood deck on steel beams or masonry-bearing walls for a steel-spanning floor or roof structure. In this book I deal primarily with steel structures, but I also discuss some of the common mixed-material situations.

2.7 CONNECTION METHODS

Nowadays structural steel members that consist of rolled elements typically are connected by direct welding or by steel bolts. (Riveting is practically obsolete.) In general, welding is preferred for shop fabrication and bolting for field connections. I discuss the design of simple welded connections and bolted connections in Chapter 11.

Thin cold-formed steel elements may be attached by welding, bolts, or sheet metal screws. Thin deck and wall paneling elements sometimes are attached by simple interlocking at their abutting edges; the interlocked parts may be folded or crimped to enhance the connection.

You can use adhesives or sealants to seal joints or bond thin sheet materials in laminated fabrications. Although some connecting elements are attached to connected parts by adhesion, easing the work of fabrication and erection, adhesion is not used for major structural joints.

Connecting columns and beams in multistory buildings is a major structural design problem. For rigid-frame action to resist lateral loads, these connections need to be very large welds to transfer the full strength of connected members. How to design these connections is beyond the scope of this book, but I discuss lighter framing connections in Chapter 11.

2.8   DATA FOR STEEL PRODUCTS

For general information regarding steel products used for building structures, check steel industry publications. For example, the AISC is the primary source of design information regarding structural rolled products, which are the principal elements used for major structural components such as columns, beams, and large trusses. Several other industry-wide organizations publish documents that provide information about particular products, such as manufactured trusses (open-web joists), cold-formed sections, and formed sheet steel decks. I describe many of these organizations and their publications elsewhere in this book as appropriate.

Manufacturers of steel products usually conform to industry-wide standards, but they often find room for variation. As a result, be sure to look up specific data and details of actual products in the manufacturers' own publications.

Designers should strive not to predetermine building components; stay flexible, specifying only the critical controls.

Some of the data I present has been reproduced or abstracted from industry publications. In many cases the data is abbreviated and limited to uses pertinent to the example computations and exercise problems. Consult the reference sources cited for more complete information, especially since change occurs frequently, reflecting technological advances, new research, and modified codes and industry standards.

3

WHAT YOU SHOULD
KNOW ABOUT STEEL

Steel is a relatively expensive product. So before choosing to use steel for structures, carefully consider steel's limitations. In addition, because steel is an industrialized product, design with a practical eye; in other words, understand the complications inherent in producing and then building with steel products. In this chapter I detail a number of aspects you should consider before using steel for structural applications.

3.1 STRESS AND STRAIN

Steel is one of the strongest materials used for building structures, but it still has stress limits. For example, unlike wood, steel has a nondirectional stress response. And unlike concrete or masonry, steel has a high stress resistance for the basic stresses (tension, compression, and shear). Some specific steel stress limits are the following:

- Stress beyond the yield point produces permanent deformations. Though tolerable within a joint, such deformations create major problems within the general form of structural elements. As a result, though steel's ultimate strength is high, you must consider the much lower yield stress as a limit of acceptable behavior in most situations.

- Ordinary steel is formed by molten casting, resulting in a crystalline structure. Fractures along crystalline fault lines can cause stress failure, especially that related to dynamic, repetitive force actions. This limitation is more of a problem in machinery, but dynamically loaded building structures may need consideration for this effect.
- Various actions, such as cold-forming, machining, or welding, can change steel's character, resulting in hardening, loss of ductility, or locked-in stresses. Before using steel, be sure the production processes do not undermine steel's stress behavior under service load situations.

In some cases the anticipated stress-strain responses may affect how you compute a steel structure's overall resistance. For example, when a plastic hinge—in a rigid frame or in a frame with eccentric bracing—yields, you must take into account a structure's adjusted behavior in load response. I discuss this case in Chapters 7 and 13.

When sharing loads with other materials, steel tends to carry a disproportionately high share because of its stiffness. This behavior is a major factor in the design of composite structural elements, such as flitched beams of steel and wood or composite deck systems of steel and concrete (see Section 9.2). This behavior also plays a part in the design of reinforced concrete and masonry.

Although stress resistance varies, strain resistance—the direct stress modulus of elasticity—does not. As a result, though higher grades of steel boast increased load resistance (as measured by stress capacity), they cannot resist deformations more. Thus deflection and buckling, which are affected by a material's stiffness, become relatively more critical for structures made with higher grades of steel.

3.2 STABILITY

Load resistance, measured in terms of strength, is the most critical limit for a structure. However, a loss of stability may lead to failure before stress levels achieve limiting magnitudes. To truly establish safety, designers therefore must consider strength and stability. Stability poses a concern because steel elements often consist of relatively thin parts, not solid forms as is common with timber and concrete. In addition, framed assemblages often consist of fairly slender linear components. This thin character means that failure frequently occurs as buckling collapse, rather than crushing or tension cracking. As a result, designers must pay special attention to the potential for various types of buckling failure. I describe these forms of failure elsewhere in this book as they pertain to certain structural elements and systems.

Another stability problem: Most assemblages do not derive much stability from their connections, which have so little moment-resistance that they often qualify as pinned connections rather than fixed connections. (In truth the typical connection is partially fixed, with some limited moment-transmitting capacity.)

Such a problem often relates to the resistance of lateral loads. However, the problem may be more general: how to give a structure some degree of three-dimensional stability. Possible soluions include

- Modify the usual connections to more fully resist moments, producing *rigid-frame* action.
- Arrange the frame so that the overall assemblage provides more stability but without rigid joints (as with triangulation that produces truss action).
- Add extra bracing elements (guys, struts, X-bracing, flying buttresses, etc.).
- Borrow stability from other building construction parts (e.g., masonry walls).

The bottom line: Lateral loads require more planning than gravity loads. In other words, designers must pay attention to what stabilizes an assembled structure. I discuss specific situations regarding stability elsewhere in the book as appropriate.

3.3 DEFORMATION LIMITS AND CONTROLS

A practical limitation is the amount of deformation steel can tolerate. *Deformation*—literally, shape change—is inevitable because stress resistance cannot be developed without some accompanying strain.

For structural members the most critical deformations are usually those caused by bending, due simply to the larger deformation dimensions. A heavily loaded column may shorten by a virtually imperceptible amount, but even a short span beam deflects noticeably when loaded. In fact, the most common deformation problem is the vertical deflection of beams.

To determine practical deflection limit—that is, the tolerable movement of a structure—consider how deflection affects the general building construction. For example, be sure that tiled floors or plastered ceilings won't crack. Although the need to control deformation is easily understood, the practical means for establishing design criteria is elusive, requiring much professional judgment. These problems are treated in some depth in the chapters that deal with spanning structures.

Note: Another deformation problem occurs within structural connections, where strain and deformation unavoidably accompany stress. I discuss some pertinent issues in Chapter 11.

3.4   RUST

When exposed to air and moisture, most steels rust at the surface. If you do nothing, the entire steel mass eventually will rust away. To prevent rusting, do one or more of the following:

- Paint the steel surface with rust-inhibiting material.
- Coat the surface with nonrusting metal, such as zinc or aluminum.
- Use a steel that contains ingredients that prevent or retard rusting (see Section 2.1 for discussion of corrosion-resistant steels).

Note: If the steel element is encased in cast concrete or other encasing construction, the steel essentially is unexposed and thus will not rust. Rusting poses a greater problem when exposure conditions are severe or when the steel elements are thin (e.g., those formed of thin sheet steel, such as formed roof decks).

Although standard building construction practices may necessitate leaving steel in an essentially bare condition, such as when field (i.e., on-site) welding is required or when the steel must be encased in concrete, designers must take special care when appearance is important—that is, when the final structure includes exposed steel.

Designers tend to avoid using excessively thin parts for structures exposed to conditions likely to cause serious rusting. This practice reduces a structure's vulnerability to failure from rust. (In other words, if prevention methods are not totally successful, the structure can afford a loss of material in cross sections.)

Steel also can deteriorate when exposed to corrosive chemicals, such as acid rain, salt air, or air heavily polluted with industrial wastes. Special protection may be required for such conditions. Where actual structural safety is at risk, not just appearance, these matters require serious attention by the structural designer.

3.5   FIRE

Steel's stress and strain response varies with temperature. At high temperatures, steel rapidly losses strength. In addition, steel loses stiffness, which may be more important when buckling is critical. Thanks to steel's high conductivity, which leads to rapid heat gain, and cost, which leads designers to use thin parts, such loss of strength makes steel structures highly susceptible to fire. On the other hand, because steel is noncombustible, the concern is less critical than for buildings constructed with thin wood elements.

The chief strategy for improving fire safety is to prevent the fire from reaching the steel by providing some coating or encasement with fire-resistant, insulative materials, including masonry, plaster, mineral fiber, gypsum plasterboard, or concrete. I discuss the problem and detail some specific design situations in Chapter 12.

Designers often use concrete as a fill on top of formed steel deck or as a slab bearing directly on steel beams (see the steel-beam-plus-concrete-slab construction shown in Figure 3.1). When designing such a system, however, you must consider weight of the concrete.

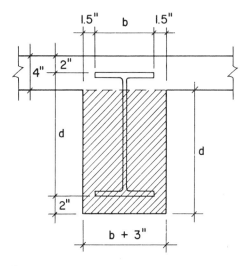

FIGURE 3.1 Steel beam encased in concrete—for fire protection. Shaded portion indicates concrete in excess of that required for the supported slab.

3.6 ASSEMBLAGE

Steel structures consist of many parts. Assembling a complete structure—that is, bringing together all the parts and connecting them—is never easy. In fact, designers must design individual parts while keeping in mind the assemblage of the complete structure. Designers must balance the following:

- *The Structural Arrangement.* Designers must decide the structure's overall form, as well as span dimensions, story heights, sizes of openings, and so on. Other decisions involve repetitive modules; for example, designers must define the spacing for sets of beams and columns.

- *How the Parts Mate.* Because decks must be attached to beams, beams attached to columns, columns attached to footings, and so on, designers must consider the structure's geometry: the shapes of individual parts affect how they are connected, which determines how loads are transferred.

Standard connecting methods relate to the forms of the connected elements and to the type of loads transferred between elements. For a typical steel structure, connecting methods should be practical and economical—and above all, familiar to the assemblage crew.

To some degree, assemblage that is performed in the factory differs from that performed at the job site. As a result part of the designer's task is to visualize where the assemblage occurs, for the assemblage site may well affect which members and connecting methods are best for the structure.

I outline assemblage problems for individual elements throughout the book. I discuss overall assemblage problems in Chapter 4. (Also refer to the building case examples in Chapter 12.)

3.7 COST

Steel is relatively expensive, on a volume basis. The real concern, however, is the final *installed cost*—that is, the total cost for the erected structure—which includes the costs of transporting components to the site and using auxiliary devices (such as connecting elements for structural components and bridging for joists) to complete the structure. Inevitably designers try to use the least volume of steel, but such economy is limited to the design of a single type of item; no matter what, rolled structural shapes do not cost the same per pound as fabricated open-web joists.

I discuss cost concerns for whole structures in Chapter 12. In general, however, when designing a single structural component, designers strive to use the lightest elements (least volume of material) that satisfy the design criteria.

4

CHOOSING A STRUCTURAL SYSTEM

Designers use steel elements in a variety of horizontal-spanning floor or roof structures. The primary spanning systems I cover in this book are the rolled steel beam and the light, prefabricated truss.

In this chapter I outline some general issues surrounding the development of spanning systems. I treat the design of beams and decks in Chapter 5 and the design of trusses in Chapter 9. I discuss rigid frames and bents in Chapter 8.

4.1 DECK-BEAM-GIRDER SYSTEMS

Structural Concerns

A framing system used extensively for buildings with large roof or floor areas consists of columns arranged in orderly rows to support a rectangular grid of steel beams or trusses. The actual roof or floor surface is then generated by a solid deck of wood, steel, or concrete, which spans in multiple, continuous spans over a parallel set of supports. When planning such a system—developing its layout and choosing its components—designers must consider the following:

> *Deck Span.* Determines the deck's general type and specific variation (thickness of plywood, gage of steel sheet, etc.).
> *Joist Spacing.* Determines the deck span. Affects the load magnitude on the joist. The joist type selected may limit the spacing, based on the joist capacity. *Note:* You must coordinate the joist type and spacing with the deck type.
> *Beam Span.* For systems with some plan regularity, the joist spacing is some full-number division of the beam span.
> *Column Spacing.* Determines the beam and joist spans; related to all other components.

When planning a system such as that shown in Figure 4.1*a*, first locate the system supports, usually columns or bearing walls. The character of the spanning system is closely related to the magnitude of the spans it must achieve. For example, decks usually have short spans, so the elements that provide direct support must sit close together. Joists and beams, meanwhile, may be small or large, depending on their spans; the larger they are, the less likely they are closely spaced. Very long span systems may have several levels of components, ranging in size from the elements that achieve the longest span down to the elements that directly support the deck.

I discuss components of the deck-beam-girder system in Chapter 6. (See also the building system design examples in Chapter 12.)

Figure 4.1*b* is a plan and elevation of a system that uses trusses for the major span. If the trusses are very large and the purlin spans quite long, the purlins often are widely spaced. Because purlins usually must coincide with the joints in the top of the truss to prevent high shear and bending in the truss top chord, it may be advisable to use joists between widely spaced purlins to provide support for the deck. On the other hand, if the truss spacing is modest, it may be possible to use a long-span deck with no purlins. As you can see, the system's basic nature changes, depending on lengths of spans and locations of supports. In any event, the truss span and

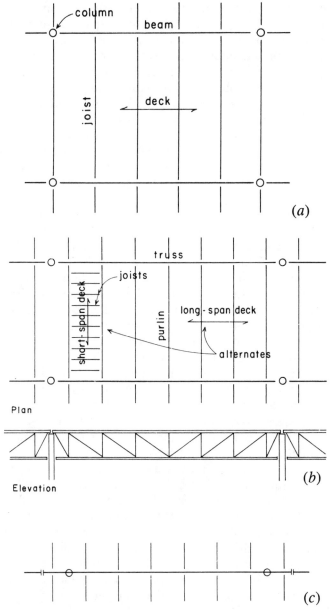

FIGURE 4.1 Beam framing—planning considerations.

panel module, the column spacing, the purlin span and spacing, the joist span and spacing, and the deck span are interrelated; selecting components is a highly interactive exercise.

For systems with multiple elements, you must consider how the components meet and connect. Figure 4.1*a* shows a five-member intersection at the column, involving the column, the two beams, and the two joists (throw in an upper column if the building is multistory). Whether this inersection is a routine matter or a mess depends on the materials and forms of the members, the forms of connections, and the types of force transfer at the joint. You can bring about some relief of the traffic by using the plan layout in Figure 4.1*c*, in which the module of the joist spacing is offset at the columns, leaving only the column and beam connections at the column location; in addition, the beam is continuous through the column, with the beam splice occurring off the column. In Figure 4.1*c*, all the connections are two-member relationships only: column to beam, beam to beam, and beam to joist.

You must also plan bridging, blocking, and cross-bracing for trusses with care because these members may interfere with continuous piping or ducting. In addition, their connections may prove problematic. To ease the congestion, consider using required bracing elements for multiple purposes. For example, blocking required for plywood nailing may function also as edge nailing for ceiling panels and as lateral bracing for slender joists or rafters. And the cross-bracing required to brace tall trusses may support ceilings, ducts, building equipment, catwalks, and so on.

Cantilevered edges, extensions of a horizontal structure beyond the plane of its exterior walls, (for example, an overhanging roof), pose another structural problem. Figure 4.2*a* shows one possible solution: simply extend the ends of joists or rafters that are perpendicular to the wall. With steel framing, this type of cantilever is achieved most easily if the extended members rest atop the supporting beam or bearing wall at the wall plane. Figure 4.2*b* is an alternate solution, for an exterior column system: extend the column-line members to support a member at the cantilevered edge, which in turn supports simple-span members between the column lines. Which scheme you choose depends on loading, member size and type, and cantilever magnitude.

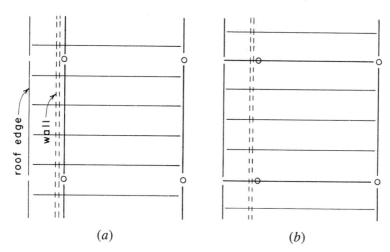

(a)　　　　　　　　　　　　*(b)*

FIGURE 4.2　Framing of cantilevered edges.

The cantilevered edge at an outside corner when both sides of the building have the cantilever condition (see Figure 4.3a) requires a special sollution. Given the framing system shown in Figure 4.2a, Figure 4.3b is a possibility: the supported beam is cantilevered to support edge member 1, and the joists are cantilevered to support edge member 2. For the system shown in Figure 4.2b, Figure 4.3c is a solution: the column-line member is cantilevered as usual to support edge member 1, which in turn cantilevers to the corner to support edge member 2. A third solution: use a diagonal member (see Figure 4.3d), which provides for a reorientation of the framing system as the corner is turned. This layout, which is more common in wood structures than in steel structures, is often used for sloping roofs when the diagonal member defines a ridge as the roof slopes to both edges. Note the busy intersection at the interior column.

The bottom line: Designers must balance structural planning with the building's general architectural design. In the end, pragmatic concerns may outweigh a theoretical optimization of a structure and its various subsystems.

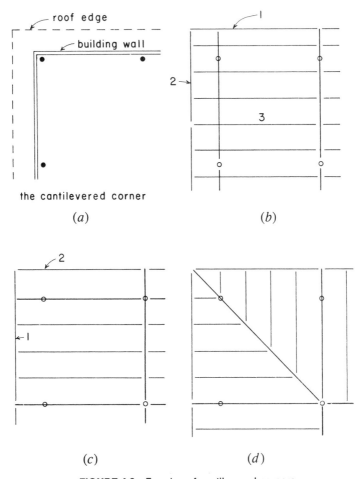

FIGURE 4.3 Framing of cantilevered corners.

General Service Concerns

I discuss various general planning concerns for structures in Chapter 12. The following are some particular issues that relate to the design of steel framing systems.

Ceilings.   In general, ceilings take one of three forms:

- Direct attachment to the overhead structure (i.e., underside of the roof or floor above)
- An independent structure achieving its own span
- Suspension from the overhead structure

Suspended ceilings are quite common, as the space between the ceiling and the structure can conceal HVAC ducts and registers, lighting system wiring and recessed fixtures, and so on. In addition, a suspended ceiling is used when the ceiling's form does not match that of the overhead structure (e.g., sloped rafters with a horizontal ceiling). Typically, ceilings are suspended either from closely spaced joists or rafters (4 ft center to center or less) or hangers attached to the deck. If you use the latter method, you free the modules of the spanning structure and the ceiling framing from each other.

Roof Drainage.   With a roof framing system, providing for the minimum slopes required to drain flat roofs is always a problem. The most direct solution is to tilt the framing to provide the slope patterns required. This approach gets quite complicated when the roof is complex: the desired slope patterns and drain locations may not relate well to the layout of various roof framing members. Another solution is to keep the framing flat but vary the deck thickness (applicable only to cast-in-place concrete decks). A third possibility is to use tapered insulation fill on top of the deck. Although the last technique simplifies the framing details, you usually can develop only a few inches slope differential. If a flat ceiling is required and is to be attached directly to the roof structure, you must consider whether a few inches slope is enough for adequate drainage.

For some types of structural members—most notably, manufactured trusses—you can slope the top of the member while keeping the bottom flat. As a result, you can have a sloping roof surface and a flat ceiling, with both the roof deck and the ceiling surfacing directly attached to the truss chords.

Dynamic Behavior. Lightweight roof structures result in dead load reduction for the spanning structure and its supports without any major drawbacks. Lightweight floor structures, on the other hand, tend to be bouncy. (*Note:* Bounciness also can result from an excessive span-to-depth ratio for the spanning elements.) Some general rules for reducing bounciness are the following:

1. Restrict live-load deflections to less than $\frac{1}{360}$ of the span for any floor.
2. Keep span-to-depth ratios well below the maximum permitted. Suggestions: maximum of 20 for rolled shapes, 15 for open web joists.
3. Do not use decks for the longest spans listed in design data references. (Result: stiffer decks.)
4. Using concrete fill on top of decks (steel or wood) also reduces bounciness.

Holes. Floor and roof surfaces often are pierced by a number of passages: large openings for stairs and elevators, medium-size openings for ducts and chimneys, and small openings for piping and wiring. Designers must plan the structure in detail to accommodate these openings. Such planning entails the following considerations:

Location. Some openings that occur at inconvenient locations are unavoidable. Others are the result of poor planning. For structures that use column-line rigid-frame bents for lateral bracing, to maintain the integrity of the bents, keep the openings off the column lines. For regularly spaced systems, coordinate the framing layout with the locations of required openings to maintain a maximum regularity. Openings should not interrupt the system's major elements (i.e., large trusses or girders).

Size. Large openings require some framing around their perimeters (such openings often coincide with locations of supported wall construction). Small openings (e.g., for single pipes), however, may simply pierce the deck, needing no special provision. For other-size openings, accommodation requirements depend on the form and size of the structural elements in question. For example, openings that fit between joists need little if any support, but openings that interrupt one or more joists require more difficult measures, such as doubling the joists that bracket the opening.

Columns. It is sometimes convenient to locate duct shafts or chases for piping or wiring next to a structural column. If you can do so without interrupting a major spanning member, great. If the opening must be on the column line, you may need to straddle the opening with a double-framing member of two spaced elements.

Diaphragms. Large openings may adversely affect how the floor or roof system functions as a horizontal diaphragm for lateral bracing. You may need to provide special framing or connections to develop collector functions, drag struts, or the subdivision of the diaphragm (refer to Section 12.7).

4.2 THE THREE-DIMENSIONAL FRAME

Steel elements frequently are used for two-dimensional structures—for example, floor, roof, or wall. However, steel elements also are arranged in three-dimensional systems—for example, a skeleton structure for a tower or multistory building. (In fact, skyscrapers, in the late nineteenth century, represent one of the early major uses of rolled shapes.)

Product development and application often grow interactively; such was the case with W shapes (originally called *wide-flange shapes*). The geometry of W shapes—they have relatively wide flanges, and most of the flange surface is flat, not tapered—facilitates the assemblage of frameworks that use a common joint configuration. Figure 4.4 shows a multistory steel column continuous between two stories, with steel beams framing into it from all four horizontal directions; the W shape column (whose cross section actually is an I or H shape) is ideal to accommodate this framing. To achieve the joint shown in Figure 4.4, the column needs the following attributes:

- The flanges must be wide enough to accept the framing connection of the beam on the flange side of the column.

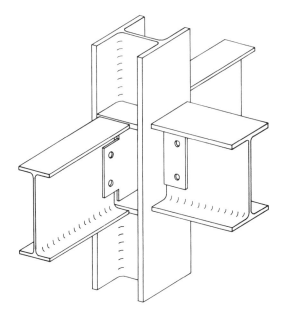

FIGURE 4.4 Detail of column-beam framing connections.

- The distance flange to flange (i.e., the *depth* of the W shape) must accommodate the framing connection of the beams that frame into the column web.
- For the given connection detail, any splice joints in the column must sit above or below the beam level.

The following are some common rules of thumb regarding the W-shape column:

Minimum Column Depth. For multistory construction, use columns at least 10 in. deep. For heavier loads and larger beam framing, depths may be 12 or 14 in.

Minimum Flange Width. Use shapes with flanges at least 6 in. wide. Wider flanges are available, of course; in fact, an approximately square column shape is common for column use in the 10, 12, and 14 in. nominal shapes series.

Common Column Splice Location. Locate the column splice about 3 ft above the tops of the beams, which is a handy height for the steel erection crew.

When planning three-dimensional frames, you must factor in all details pertinent to two-dimensional systems. In addition, you should consider the following:

- Columns should be located over columns in the story below whenever possible.
- Framed bents may be constituted by the columns and beams in a single vertical plane.

(I discuss the nature and problems of such bents in Chapter 8.)

4.3 TRUSS SYSTEMS

Using trussing produces very light steel structures. Two basic principles justify the use of trusses:

- A triangle's basic planar stability; it is held rigid simply because its sides resist a change in length.
- The isolation of highly efficient concentrations of material; such great separation is possible only with a relatively strong material.

I analyze the usage of steel trusses in Chapter 10.

In some steel frameworks, trusses replace beams—for example, to achieve spans more efficiently, especially when spans are great, or to ease the passing of ducting, piping, or wiring through the system.

Adding diagonal members to a vertical planar arrangement of steel beams and columns produces a *trussed bent*. This use of trussing is common to give steel frameworks three-dimensional stability; in fact, the trussed bent is a common bracing option developed to resist the horizontal force effects of wind or earthquakes.

Trussing is highly adaptable, appropriate for many forms other than rectilinear frameworks. Linear rolled shapes can be bent or curved, but trussing can be shaped easily to achieve just about any form.

4.4   RIGID FRAMES

Rigid frames are frameworks that use rigid (i.e., moment-resistive) connections. Normal connections involve common connecting devices; rigid connections require specially developed *joints*. Figure 4.4, for example, shows the connection between a beam and supporting column: a connecting device is attached to the beam web and then to the column. This kind of connection essentially transfers only vertical load. To transmit bending moment to the column, you must connect the beam flanges to the column, thereby producing a rigid joint.

One way to achieve a rigid joint is to weld beam flanges directly to a column. In that case the beams on the column's flange side enjoy a direct connection, achieved with a butt weld at the end of the beam flange. Because bending moment in the column is developed most effectively by the column flanges, this connection effects reasonably direct transfer between beam and column. However, since the beam grabs only one column flange, you should enhance the joint by welding filler plates (see Figure 4.4) to the inside of the column at the level of the beam flanges. Doing so helps transfer the bending across the whole column section.

To achieve the most effective column-beam rigid-frame bent, turn the W shape columns so that the column flanges are perpendicular to the plane of the planar framed bent. In some situations, however, you may need to achieve bent action in both directions; to do so, attach the beams that intersect the open side of the columns for moment transfer—not an easy task.

I describe various connection problems in Chapter 11. I discuss rigid frames in general in Chapter 8 and rigid-frame bents specifically in Sections 12.3, 12.5, and 12.7.

4.5   MIXED SYSTEMS

The all-steel structure is possible, but the typical building consists of a variety of materials. For example, steel spanning roof or floor systems are supported sometimes by steel columns but frequently by concrete or masonry walls. Steel frameworks are surfaced sometimes with formed sheet steel but frequently with plywood panels, cast-in-place concrete decks, or precast concrete panels.

When planning a mixed-material system, you must take into account the problems of all the materials and structural elements involved. This book is essentially about steel structures, but I touch on some issues regarding mixed-material structures in Chapter 6, when I discuss supports for steel spanning elements. Also refer to the case study examples in Chapter 12.

5

STRUCTURAL
INVESTIGATION AND DESIGN

To plan any structure, you must use structural investigation and design methods.

Essentially an analytical activity, *investigation* is the identification of how a structure behaves when subjected to some loading condition. When a structure is unique, investigation also may involve discovery, evaluation, or both. *Design* is the overall decision-making process that leads to a fully defined structure. For example, defining a structure is a design process; it anticipates an analytical investigation.

In this chapter I describe structural investigation and design as they relate currently to the design of steel structures for buildings.

5.1 INVESTIGATION OF STRUCTURAL BEHAVIOR

Investigating how structures behave is an important part of structural design: it provides a basis for ensuring the adequacy and safety of a design. In this section I discuss structural investigation in general. As I do throughout this book, I focus on material relevant to structural design tasks.

Purpose of Investigation

Most structures exist because they are needed. Any evaluation of a structure thus must begin with an analysis of how effectively the structure meets the usage requirements.

Designers must consider the following three factors:

- *Functionality*, or the general physical relationships of the structure's form, detail, durability, fire resistance, deformation resistance, and so on.
- *Feasibility*, including cost, availability of materials and products, and practicality of construction.
- *Safety*, or capacity to resist anticipated loads.

Means

An investigation of a fully defined structure involves the following:

1. Determine the structure's physical being—materials, form, scale, orientation, location, support conditions, and internal character and detail.
2. Determine the demands placed on the structure—that is, loads.
3. Determine the structure's deformation limits.
4. Determine the structure's load response—how it handles internal forces and stresses and significant deformations.
5. Evaluate whether the structure can safely handle the required structural tasks.

Investigation may take several forms. You can

- Visualize graphically the structure's deformation under load.
- Manipulate mathematical models.

- Test the structure or a scaled model, measuring its responses to loads.

When precise quantitative evaluations are required, use mathematical models based on reliable theories or directly measure physical responses. Ordinarily, mathematical modeling precedes any actual construction—even of a test model. Limit direct measurement to experimental studies or to verifying untested theories or design methods.

Visual Aids

In this book, I emphasize graphical visualization; sketches are invaluable learning and problem-solving aids. Three types of graphics are most useful: the free-body diagram, the exaggerated profile of a load-deformed structure, and the scaled plot.

A *free-body diagram* combines a picture of an isolated physical element with representations of all external forces. The isolated element may be a whole structure or some part of it.

For example, Figure 5.1a shows an entire structure—a beam-and-column rigid bent—and the external forces (represented by arrows), which include gravity, wind, and the reactive resistance of the supports (called the *reactions*). *Note:* Such a force system holds the structure in static equilibrium.

Figure 5.1b is a free-body diagram of a single beam from the bent. Operating on the beam are two forces: its own weight and the interaction between the beam ends and the columns to which the beam is attached. These interactions are not visible in the free-body diagram of the whole bent, so one purpose of the diagram for the beam is to illustrate these interactions. For example, note that the columns transmit to the ends of the beams horizontal and vertical forces as well as rotational bending actions.

Figure 5.1c shows an isolated portion of the beam length, illustrating the beam's internal force actions. Operating on this free body are its own weight and the actions of the beam segments on the opposite sides of the slicing planes, since it is these actions that hold the removed portion in place in the whole beam.

Figure 5.1d, a tiny segment, or particle, of the beam material is isolated, illustrating the interactions between this particle and

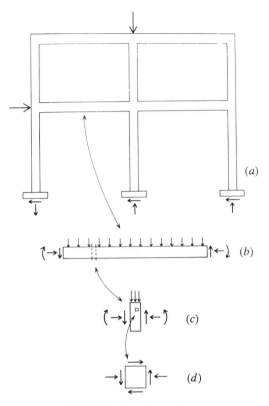

FIGURE 5.1 Free-body diagrams.

those adjacent to it. This device helps designers visualize stress; in this case, due to its location in the beam, the particle is subjected to a combination of shear and linear compression stresses.

An *exaggerated profile of a load-deformed structure* helps establish the qualitative nature of the relationships between force actions and shape changes. Indeed, you can infer the form deformation from the type of force or stress, and vice versa.

For example, Figure 5.2a shows the exaggerated deformation of the bent in Figure 5.1 under wind loading. Note how you can determine the nature of bending action in each member of the frame from this figure. Figure 5.2b shows the nature of deformation of individual particles under various types of stress.

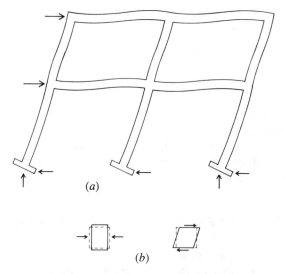

(a)

(b)

FIGURE 5.2   Structural deformation.

The scaled plot is a graph of some mathematical relationship or real data. For example, the graph in Figure 5.3 represents the form of a damped vibration of an elastic spring. It consists of the plot of

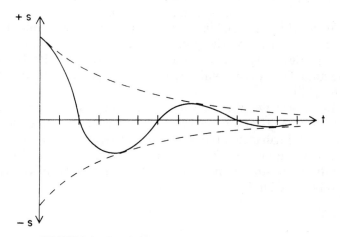

FIGURE 5.3   Graphical plot of a damped cyclic motion.

the displacement s against elapsed time t, and represents the graph of the expression

$$s = \frac{1}{e^t} P \sin(Qt + R)$$

Although the equation is technically sufficient to describe the phenomenon, the graph illustrates many aspects of the relationship, such as the rate of decay of the displacement, the interval of the vibration, the specific position at some specific elapsed time, and so on.

5.2 METHODS OF INVESTIGATION AND DESIGN

Traditional structural design centered on the *working stress method*, a method now referred to as *stress design* or *allowable stress design* (ASD). This method, which relies on the classic theories of elastic behavior, measures a design's safety against two limits: an acceptable maximum stress (called *allowable working stress*) and a tolerable extent of deformation (deflection, stretch, etc.). These limits refer to a structure's response to *service loads*—that is, the loads caused by normal usage conditions. The *strength method*, meanwhile, measures a design's adequacy against its absolute load limit—that is, when the structure must fail.

To convincingly establish stress, strain, and failure limits, tests were performed extensively in the field (on real structures) and laboratories (on specimen prototypes, or *models*). *Note:* Real-world structural failures are studied both for research sake and to establish liability.

In essence, the working stress method consists of designing a structure to *work* at some established percentage of its total capacity. The strength method consists of designing a structure to *fail*, but at a load condition well beyond what it should experience. Clearly the stress and strength methods are different, but the difference is mostly procedural.

The Stress Method (ASD)

The stress method is as follows:

1. Visualize and quantify the service (working) load conditions as intelligently as possible. You can make adjustments by determining statistically likely load combinations (i.e., dead load plus live load plus wind load), considering load duration, and so on.
2. Establish standard stress, stability, and deformation limits for the various structural responses—in tension, bending, shear, buckling, deflection, and so on.
3. Evaluate the structure's response.

An advantage of working with the stress method is that you focus on the usage condition (real or anticipated). The principal disadvantage comes from your forced detachment from real failure conditions—most structures develop much different forms of stress and strain as they approach their failure limits.

The Strength Method (LRFD)

The strength method is as follows:

1. Quantify the service loads. Then multiply them by an adjustment factor (essentially a safety factor) to produce the *factored load*.
2. Visualize the various structural responses and quantify the structure's ultimate (maximum, failure) resistance in appropriate terms (resistance to compression, buckling, bending, etc.). Sometimes this resistance is subject to an adjustment factor, called the *resistance factor*. When you employ load and resistance factors, the strength method is now sometimes called *load and resistance factor design* (LRFD) (see Section 5.9).

3. Compare the usable resistance of the structure to the ultimate resistance required (an investigation procedure), or a structure with an appropriate resistance is proposed (a design procedure).

A major reason designers favor the strength method is that structural failure is relatively easy to test. What is an appropriate working condition is speculation. In any event, the strength method which was first developed for the design of reinforced concrete structures, is now largely preferred in all professional design work.

Nevertheless, the classic theories of elastic behavior still serve as a basis for visualizing how structures work. But ultimate responses usually vary from the classic responses, because of inelastic materials, secondary effects, multimode responses, and so on. In other words, the usual procedure is to first consider a classic, elastic response, and then to observe (or speculate about) what happens as failure limits are approached.

5.3   INVESTIGATION OF COLUMNS AND BEAMS

Structural investigation begins with an analysis of the entire structure to determine the types and magnitudes of interior force actions (tension, compression, shear, bending, and torsion), as well as responses at supports (i.e., reactions). For systems that consist of simple beams and individual, pin-ended columns (e.g., wood frames), this analysis is quite easy. Most concrete frame structures, on the other hand, not only have members that are continuous through many spans, but also contain beam-and-column groups that constitute rigid frames. Concrete frames are thus often statically indeterminate and their investigation is complex.

Most steel frames fall somewhere between the simple and the complex. Although most steel beams are simple-spanning and most one-story columns are typically pin-ended, beams sometimes are made continuous and multistory columns sometimes are extended as a single piece through more than one story. In special situations, designers may call for a steel frame with moment-resistive joints—for example, to resist lateral forces from wind or earthquakes.

Analyzing complex indeterminate structures is beyond the scope of this book, but I attempt in this chapter to explain some aspects of their behavior. *Note:* For an approximate design, you may use any approximation method (see Section 5.5).

Investigation of Columns

Investigating columns begins with a concern for direct, axial compression. If the column is slender, you also must investigate for buckling. In rigid-frame structures, columns also are subjected to bending and shear; the commonly used three-dimensional rigid frame typically endures bending in two directions and occasionally torsional twisting. Furthermore, because precise construction of columns is difficult, designers assume bending will occur, even for simple, axially loaded columns.

I discuss the analysis and design of columns in Chapter 7. I detail some aspects of column-beam frame behavior in Chapter 8.

Investigation of Beams

The simple beam (single-span beam) rarely occurs in reinforced concrete structures, existing when a single span is supported on bearing-type supports that offer little restraint (see Figure 5.4*a*) or when beams are connected to columns with connections that offer little moment resistance (Figure 5.4*b*). However, the simple beam is common in wood and steel structures.

Single-story structures supported on bearing-type supports can experience complex bending when the spanning members are extended over the supports, in the form of cantilevered ends (Figure 5.4*c*) or multiple spans (Figure 5.4*d*). These conditions are common in wood and steel structures and occasionally crop up in reinforced concrete structures. Wood and steel members usually are constant in cross section throughout their length so designers need only find the single maximum value for shear and the single maximum value for moment. Figure 5.5 shows common conditions in concrete structures when beams and columns are cast monolithically. For the single-story structure (Figure 5.5*a*), the rigid joint between the beam and its supporting columns causes the behavior shown in Figure 5.5*b*—the columns offer some degree of

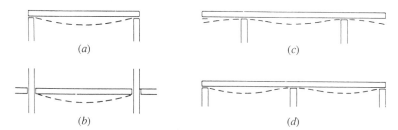

FIGURE 5.4 Flexural deformation of various beam forms.

restraint to the rotation of the beam ends. Thus some moment is added to the tops of the columns and the beam behaves like the center portion of the span in Figure 5.4c, with both positive and negative moments.

For the multistory, multispan rigid frame, the typical behavior is as shown in Figure 5.5d—the columns above and below, plus the beams in adjacent spans, contribute to the development of restraint for the ends of an individual beam span. This condition is normal in concrete structures but occurs in steel structures only when welded or heavily bolted moment-resisting connections are used.

The structures in Figures 5.4d, 5.5a, and 5.5c are statically indeterminate—that is, you cannot investigate them using only the conditions of static equilibrium. Although a detailed discussion of

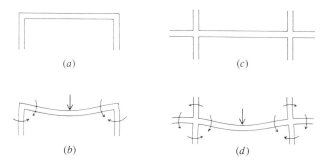

FIGURE 5.5 Beam actions in rigid frames.

statically indeterminate behaviors is beyond the scope of this book, I describe the factors affecting the behavior of continuous frames, and I provide material for approximate analysis of common situations.

Effects of Beam End Restraint. Figures 5.6*a* to *d* illustrate the effects of various end support conditions on a single-span beam with a uniformly distributed load. Each figure includes the maximum shear and moment; end reaction forces are the same as the end shears.

Figure 5.6*a* shows a *cantilever beam*, which is supported at only one end, which is fixed. Both shear and moment are critical at the fixed end, and maximum deflection occurs at the unsupported end.

Figure 5.6*b* shows a classic simple beam, whose supports offer only vertical force resistance. This type of support is called a *free end* (meaning that it is free of rotational restraint). Shear is critical at the supports, and both moment and deflection are maximum at the span's center.

Figure 5.6*c* shows a beam with one free end and one fixed end, an unsymmetrical situation for the vertical reactions and the shear. Shear is critical at the fixed end. The maximum moment is negative (at the fixed end), and maximum deflection occurs at some point slightly closer to the free end.

Figure 5.6*d* shows a beam with both ends fixed, a symmetrical support condition that results in a symmetrical situation for the reactions, shear, and moments. The maximum deflection occurs at midspan. *Note:* The shear diagram is the same as for the simple beam.

Continuity and end restraint have positive and negative effects. The most positive gain is the reduction of deflections, which is generally more significant for steel and wood structures because deflections are rarely as critical for concrete members. For the beam with one fixed end, the maximum shear increases and the maximum moment is the same as for the simple span (no gain in either regard). For the beam with both ends fixed, the shear is unchanged, while both moment and deflection are reduced; still, in rigid frames, although restraints reduce moment and deflection for the beam, the columns then must take some moment. Designers

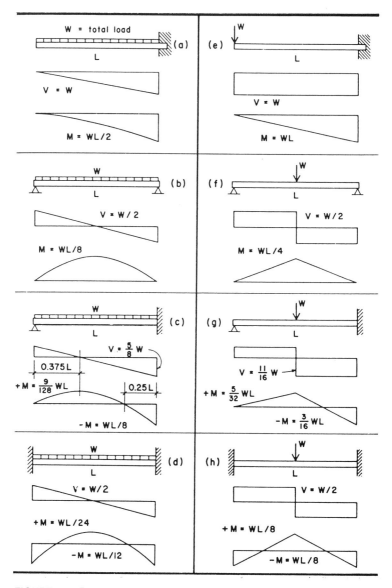

FIGURE 5.6 Response values for beams with uniformly distributed loading and single concentrated loading.

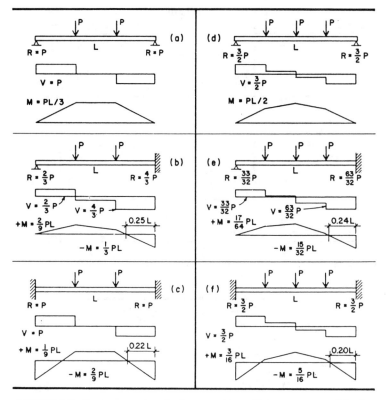

FIGURE 5.7 Response values for beams with multiple concentrated loads.

use rigid frames to resist lateral loads from wind and earthquakes; you must investigate the resulting complex combinations of lateral and gravity loading.

Effects of Concentrated Loads. Framing systems for roofs and floors often consist of evenly spaced beams supported by other beams perpendicular to them. The beams supporting beams are subjected to a series of spaced, concentrated loads—the end reactions of the supported beams. Figures 5.6e to h show the effects of a single such load at the center of a beam span. (Figure 5.7 shows two

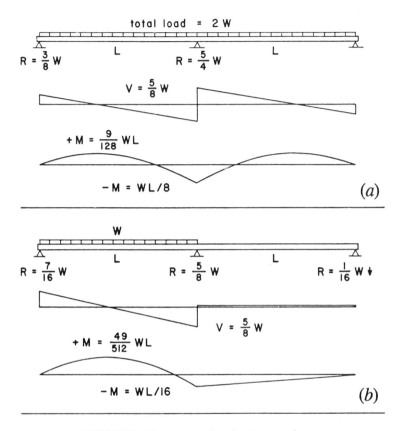

FIGURE 5.8   Response values for two-span beams.

more examples of evenly spaced concentrated loading.) *Note:* When more than three such loads occur, you may consider the sum of the concentrated loads as a uniformly distributed load and use the values given in Figures 5.6*b* to *d*.

Multiple Beam Spans.   Figure 5.8 shows various loading conditions for a beam that is continuous through two equal spans. When continuous spans occur, you usually must consider the possibilities of partial beam loading, as in Figures 5.8*b* and *d*. For example, the beam in Figure 5.8*b* has less total load but higher values for maximum positive moment and shear at the free end than the fully loaded beam in Figure 5.8*a*.

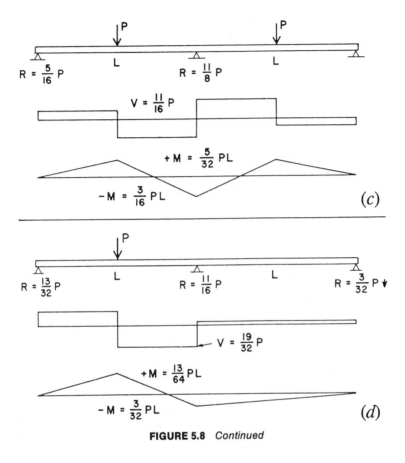

FIGURE 5.8 *Continued*

Note: Designers calculate a beam's full action by combining the partial loading effects due to *live loads* (people, furniture, snow, etc.) with those produced by *dead load* (permanent construction weight).

Figure 5.9 shows a beam continuous through three equal spans. Figure 5.9*a* gives the loading condition for dead load (*always* present in *all* spans), while Figures 5.9*b* to *d* show the partial loading possibilities, each of which produces specific critical values for reactions, shears, moments, and deflections.

Complex Loading and Span Conditions.   Figures 5.6 through 5.9 illustrate common situations. For other cases—such as unsymmetrical loadings, unequal spans, cantilevered free ends, and so

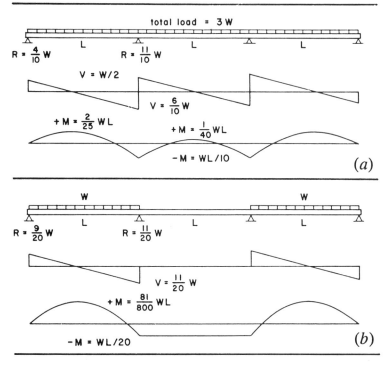

FIGURE 5.9   Response values for three-span beams.

on—designers must analyze the indeterminate structure. For additional conditions, refer to handbooks that contain tabulations similar to those presented in this book. Two such references are the *CRSI Handbook* and the AISC Manual.

I discuss loads in Chapter 12. Because structures exist to resist loads, designers must understand loads, including their sources and how reliably you can quantify them.

5.4   INVESTIGATION OF COLUMN-AND-BEAM FRAMES

Frames in which two or more members are attached to each other with connections that can transmit bending between the ends of the members are called *rigid frames*. The connections are called

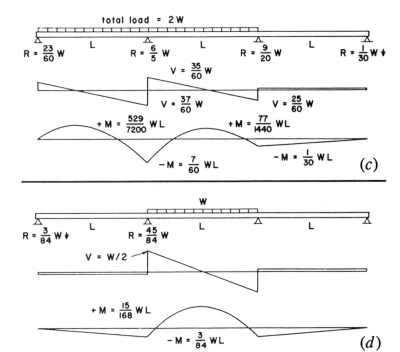

FIGURE 5.9 *Continued*

moment connections or *moment-resisting connections*. Most rigid-frame structures are *statically indeterminate*—that is, you cannot investigate them fully by considering static equilibrium alone. *Note:* The examples I present in this section, however, deal with rigid frames that are statically determinate and thus capable of being fully investigated by methods developed in this book.

Cantilever Frames

The frame in Figure 5.10a consists of two members rigidly joined at their intersection. The vertical member is fixed at its base, providing the necessary support condition for frame stability. The horizontal member is loaded with a uniformly distributed loading and functions as a simple cantilever beam. Due to the single fixed support, the frame is known as a *cantilever frame*.

Figures 5.10*b* through *f* are aids that help designers investigate a frame's behavior:

(a) The free-body diagram of the entire frame shows the loads and the components of the reactions. Studying this type aid helps designers establish the nature of reactions and determine the conditions necessary for frame stability.

(b) The free-body diagrams of individual elements help designers visualize how a frame's parts interact. They also help designers compute the internal forces in the frame.

(c) The shear diagrams of the individual elements help designers visualize or even compute the moment variations in individual elements. No particular sign convention is necessary unless to conform with the sign used for moment.

(d) The moment diagrams for the individual elements help designers defer frame deformation. The sign convention used is that of plotting the moment on the element's compression side.

(e) The deformed shape of the loaded frame is an exaggerated profile of the bent frame, usually superimposed on an outline of the unloaded frame for reference. This type aid helps designers visualize frame behavior, especially the character of external reactions and the interaction between frame parts. I find it useful to compare the deformed shape with the moment diagram.

When performing investigations, sketch the deformed shape first so that you may use it to double-check other investigative work.

Examples

The following examples illustrate the investigation of simple cantilever frames.

Example 1. Find the components of the reactions and draw the free-body, shear, and moment diagrams and the deformed shape of the frame shown in Figure 5.11*a*.

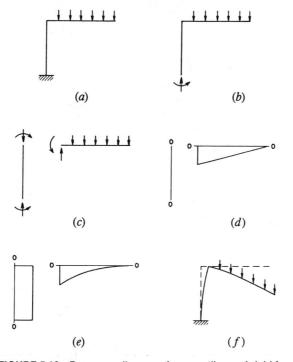

FIGURE 5.10 Response diagrams for a cantilevered rigid frame.

Solution: First determine the reactions. Considering the free-body diagram of the whole frame (Figure 5.11*b*), compute the reactions as follows:

$$\Sigma F = 0 = +8 - R_y, \quad R_y = 8 \text{ kips (up)}$$

With respect to the support,

$$\Sigma M = 0 = M_R - (8 \times 4), \quad M_R = 32 \text{ kip-ft (clockwise)}$$

Note that the *sense*, or sign, of the reaction components comes from the logical development of the free-body diagram.

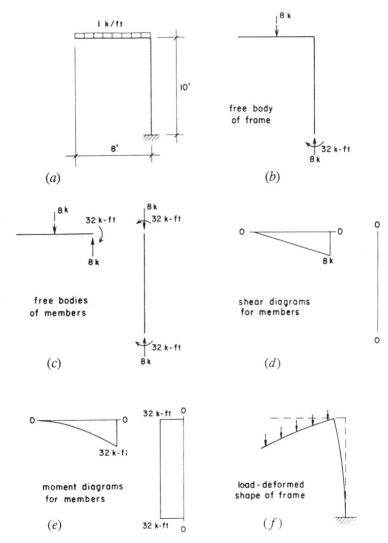

FIGURE 5.11   Behavior of the frame in Example 1.

The free-body diagrams of the individual members yield the actions that the moment connection must transmit. You can compute these actions by applying the conditions for equilibrium to either of the frame's members. Note that the force and moment senses are opposite for the two members, indicating that what one does to the other is the opposite of what is done to it.

In this example there is no shear in the vertical member. As a result, the moment from top to bottom of the member does not vary. The free-body diagram of the member, the shear and moment diagrams, and the deformed shape all should corroborate this fact. The shear and moment diagrams for the horizontal member are simply those for a cantilever beam.

With this example, as with many simple frames, you can visualize the nature of the deformed shape without resorting to any computations. I suggest such visualization as a first step in any investigation. I also urge you to check continually that individual computations conform logically to the nature of the deformed structure.

Example 2. Find the components of the reactions and draw the shear and moment diagrams and the deformed shape of the frame shown in Figure 5.12*a*.

Solution: Since the loads and reactions constitute a general coplanar force system, three reaction components are required for stability. From the free-body diagram of the whole frame (Figure 5.12*b*), use the three conditions for equilibrium for a coplanar system to find the horizontal and vertical reaction components and the moment component. If necessary, you can combine the reaction force components into a single force vector, but doing so is seldom required for design purposes.

Note that the inflection occurs in the larger vertical member because the moment of the horizontal load about the support is greater than that of the vertical load. In this case, you must compute this value before you can draw the deformed shape accurately.

Be sure to verify that the free-body diagrams of the individual members are in equilibrium and that the required correlation exists among the diagrams.

Example 3. Investigate the frame shown in Figure 5.13 for the reactions and internal conditions. Note that the right-hand support allows for an upward vertical reaction only, whereas the left-hand support allows for vertical and horizontal components. Neither support provides moment resistance.

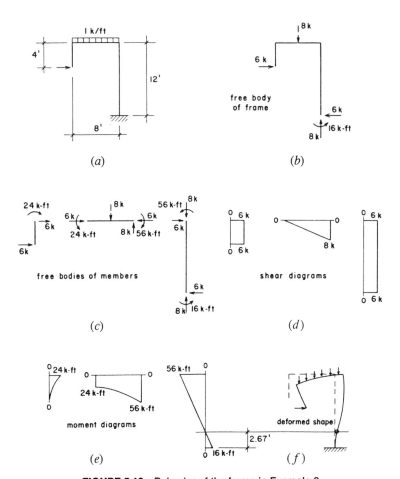

FIGURE 5.12 Behavior of the frame in Example 2.

Solution: Do the following:

1. Sketch the deflected shape (a little tricky, but a good exercise).
2. Consider the equilibrium of the free-body diagram for the whole frame to find the reactions.

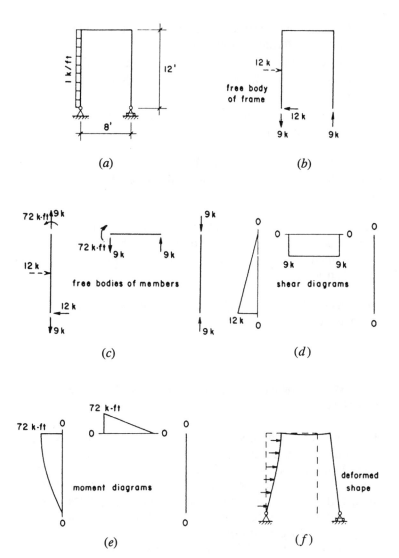

(a)

(b)

(c)

(d)

(e)

(f)

FIGURE 5.13 Behavior of the frame in Example 3.

73

3. Consider the equilibrium of the left-hand vertical member to find the internal actions at its top.
4. Consider the equilibrium of the horizontal member.
5. Consider the equilibrium of the right-hand vertical member.
6. Draw the shear and moment diagrams. Then check that the required correlation exists.

Note: Before attempting the following exercise problems, the reader should produce the results shown in Figure 5.13 independently.

Problems 5.4.A,B,C. For the frames shown in Figures 5.14*a* to *c*, find the components of the reactions, draw the free-body, shear, and moment diagrams and sketch the deformed shape of the loaded structure.

Problems 5.4.D,E. Investigate the frames shown in Figures 5.14*d* and *e* for reactions and internal conditions.

5.5 APPROXIMATE INVESTIGATION OF INDETERMINATE STRUCTURES

Investigating the highly indeterminate rigid frame, a complex structure, is a good time to use computer-aided methods; frequently used by professional designers are programs that incorporate the finite element method. Hand computation methods such as the moment distribution method were popular in the past, but they take a considerable effort—and then produce answers for only one loading condition.

Rigid-frame behavior is much simplified when the frame's joints are not displaced (that is, they move only by rotating), but

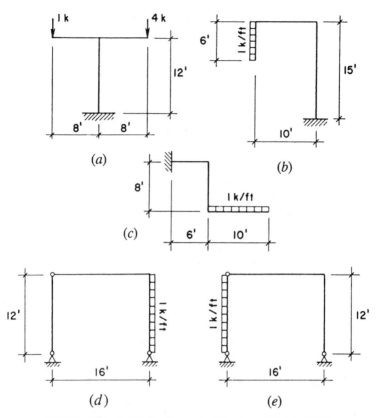

FIGURE 5.14 Reference figures for Problems 5.4.A,B,C,D,E.

such a situation usually occurs only when a symmetrical frame bears a symmetrical gravity load. If the frame is unsymmetrical, the load is nonuniformly distributed, or lateral loads are applied, frame joints move sideways (called *sidesway*) and additional forces are generated by the joint displacements.

If joint displacement is considerable, you may find force effects significantly increased in vertical members due to the P-delta effect (see Section 7.11). In stiff frames with heavy members, the change usually is not critical. In highly flexible frames (such as wood or steel frames), however, the effects can be serious. Then you must compute the actual lateral movements of the joints to obtain the eccentricities that determine the P-delta effect. Reinforced concrete frames are typically quite stiff, so this effect is often less critical than for more flexible frames of wood or steel.

Lateral deflection of a rigid frame is related to the frame's general stiffness. When several frames share a loading, as in the case of a multistory building with several bents, you must determine the frame's relative stiffnesses; to do so, consider their relative deflection resistances.

Two common types of rigid frames are the *single-span bent* and the *vertical, planar bent*, (which consists of multistory columns and multispan beams in a single plane).

The Single-Span Bent

Figure 5.15 shows two common single-span bents. The frame in Figure 5.15*a* has pinned bases for the columns, resulting in the load-deformed shape shown in Figure 5.15*c* and the reaction components as shown in Figure 5.15*e*, the free-body diagram for the whole frame. The frame in Figure 5.15*b* has fixed bases for the columns, resulting in the behavior indicated in Figures 5.15*d* and *f*. The base condition depends on the supporting structure as well as the frame itself.

The frames in Figure 5.15 are statically indeterminate and so require analysis by something more than statics. However, if the frame is symmetrical and the loading uniform, the upper joints do not move sideways and the behavior is classic; as a result, you may analyze by moment area, three-moment equation, or moment distribution, although you also can find tabulated values for behaviors for this common structural form.

Figure 5.16 shows single-span bents under a lateral load applied at the upper joint. As the upper joints move sideways, the frames change shape (Figures 5.16*a* and *b* show the reaction components).

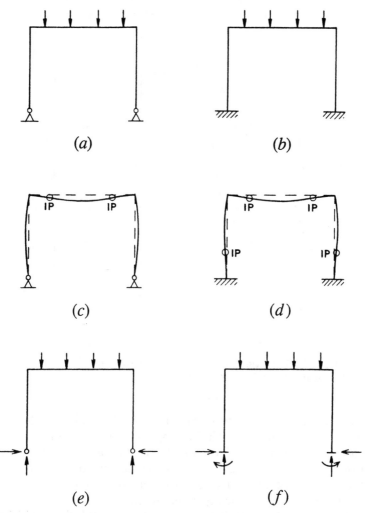

FIGURE 5.15 Responses of a rigid-framed bent with (a) pinned column base (b) fixed column base, under gravity loading.

These frames also are statically indeterminate, although some aspects of the solution may be evident. For the pinned base frame in Figure 5.16a, for example, a moment equation about one column

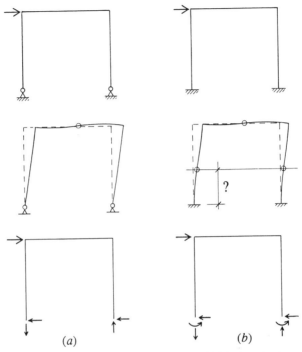

FIGURE 5.16 Responses of a rigid-framed bent with (a) pinned column base (b) fixed column base, under lateral loading.

base cancels out the vertical reaction at that location, plus the two horizontal reactions, leaving a single equation for finding the value of the other vertical reaction. Then if the bases are considered to have equal resistance, the horizontal reactions equal one-half of the load. The behavior of the frame is thus completely determined, even though it is technically indeterminate.

For the fixed column base frame in Figure 5.16b, you can use a similar procedure to find the value of the direct force components of the reactions. However, the value of the moment at the fixed base is not subject to such simplified procedures. For this investigation—as well as those for the frames in Figure 5.15—you must consider the members' relative stiffness, as you do in the moment distribution method or in any method for solution of the indeterminate structure.

Vertical Planar Bent

The rigid-frame structure occurs quite frequently as a multiple-level, multiple-span bent, constituting part of the structure for a multistory building. In most cases, such a bent serves as a lateral bracing element, although once it is formed as a moment-resistive framework, it responds as such for all types of loads.

The multistory rigid bent is also indeterminate. Its investigation is complex, requiring designers to consider several different loading combinations. When loaded or formed unsymmetrically, this bent moves sideways, further complicating attempts to analyze internal forces. Except for very early design approximations, the analysis is now sure to be done with a computer-aided system. The software for such a system is quite readily available.

For preliminary design purposes, it is sometimes possible to use approximate analysis methods to obtain member sizes of reasonable accuracy. Actually, many of the older high-rise buildings still standing were completely designed with these techniques—a reasonable testimonial to their effectiveness.

5.6 SERVICE LOAD CONDITIONS

A structure's basic task is defined essentially by its *service load conditions*, which describe what the structure is expected to resist (in service). Such a definition is an attempt to identify load sources and effects. Structural tasks are specified in terms of types (or sources) of loads, their magnitudes, and their manner of application. In addition, designers must anticipate possibile load combinations that may occur simultaneously.

Note: I discuss how designers use current references to design for loads in Section 12.1; later in that chapter I describe the need to visualize load conditions in terms of specific building usage.

How structures resist a load is a major part of structural design. Designers also try to anticipate the nature and magnitude of deformation of the structure. In many situations deformations may be tolerated by the structure but not by supported or enclosing construction (e.g., plastered ceilings, masonry walls, glass in windows, etc.). Thus deformation limits often come from various nonstructural concerns. I discuss how types of deformations affect different structural elements throughout the book; I focus on derived limits, which apply to specific building cases, in the examples in Chapter 12.

Load Types and Combinations

Loads derive from various sources, especially gravity, wind, and earthquakes. When investigating, you first must identify measure, or quantify loads in some way; when designing, you then may factor them. In most design work, you must account for all statistically likely, load combinations; doing so typically produces more than one load condition.

The *Uniform Building Code* requires you to consider the following combinations as minimum conditions for any structure:

- Dead plus floor live plus roof live (or snow)
- Dead plus floor live plus wind (or seismic)
- Dead plus floor live plus wind plus snow/2
- Dead plus floor live plus snow plus wind/2
- Dead plus floor live plus snow plus seismic

Furthermore, many structures have special concerns. For example, the stability of a shear wall may be critical with a combination of dead load and lateral load (wind or seismic). And long-term stress conditions in wood or the effects of creep in concrete may be critical with only dead load as a permanent load condition. In the end, good engineering design judgment must prevail if you are to visualize the necessary combinations.

For some structures you need consider a single load combination only. However, in complex structures with many parts (trusses, moment-resistive building frames, etc.), you must design separate members for different critical load combinations. As a result, while it may be easy to visualize the critical combination for simple structures, it is sometimes necessary to completely investigate many combinations and then compare the results in detail to ascertain true design requirements.

Factored Loads

The various load sources (gravity dead, gravity live, wind, seismic, etc.) are different in nature; in turn, they have different statistical significance. For both the stress method and the strength method, you may make some adjustments by factoring (i.e., increasing or decreasing by a factor) the individual service loads. *Note:* Factoring gets somewhat more complicated when combined loads are used.

In the stress method, load factoring is accomplished by adjusting the allowable stresses. The usual process is to give a basic allowable stress, defining "normal" conditions, and then give adjustment factors for other conditions.

As I described in Section 5.10, the strength method involves the adjusting of design loads: factors are applied individually to different types of load (dead, live, wind, etc.). To completely investigate all reasonable load combinations for a complex, indeterminate structure—for which loads may occur in several different patterns of distribution—you face a mountain of computations.

Examples of applications of factored loads are given in the work throughout this book. Factors are established by building codes (or by some reference adopted by the codes). However, designers regularly inflate factors based on personal experience.

5.7 ALLOWABLE STRESS DESIGN (ASD)

As I described in Section 5.2, the allowable stress design method (called ASD) involves limiting stresses to control structural safety. The requirements for this method are established primarily by the *Specification for Structural Steel Buildings—Allowable Stress Design and Plastic Design*, presented in the *Manual of Steel Construction—Allowable Stress Design* (called the AISC Manual—ASD for short), published by the American Institute of Steel Construction (AISC). The manual also contains basic data about standard steel products and extensive tables and graphs for design support.

The term *structural steel* refers essentially to structures produced with hot-rolled shapes (W, C, T, angles, etc.)—in other words, to major building frameworks. However, steel structures include many other types of steel elements and industrial products—for example, formed sheet steel deck, manufactured trusses (open-web joists), light-gage metals, structural bolts, and welding. Each item or group has its own standards, some of which are listed in the AISC Specification.

I use currently accepted industry standards as basic references for the work in this book, but note that all these standards are in constant flux. Designers must always keep in mind the dynamic nature of the information.

I present sample materials from various sources—mostly the AISC—throughout this book. To obtain the full range of information—and the most up-to-date information—refer to industry sources.

5.8 SERVICE CONDITIONS VERSUS LIMIT STATES

Stress methods emphasize behavior at service (i.e., actual anticipated usage) load conditions, while strength methods relate primarily to limits of resistance. Some service behavior—such as that pertaining to deformations, fire resistance, and rusting—applies to both methods. Moreover, the service load is visualized as accurately as possible for the strength method because it serves as the basis for derivation of the factored load.

Still, strength design focuses on defining a structure's behavioral limits. Two critical limits for steel are the following:

Yielding of Ductile Steel. At this stress level steel's extensive deformation precipitates various forms of failure—not necessarily a structural collapse, but perhaps a shift in failure mechanisms.

Buckling of Slender Elements. Steel elements often are slender; in addition, they often are formed with thin parts (thin webs, thin flanges, etc.). Both conditions can cause various forms of buckling that are limiting failure conditions.

Both the ASD method and the LRFD method combine elements of the old stress method with newer elements of limit states analyses. The fundamental design objective is to kill two structural birds with one stone: ensure that a structure works under service conditions and make it safe by intelligent evaluation of its limiting capacity.

5.9 LOAD AND RESISTANCE FACTOR DESIGN (LRFD)

The requirements for the LRFD method are described in the *Load and Resistance Factor Design Specification for Structural Steel Buildings*, published by the AISC. You can find information and design aids in the AISC's *Manual of Steel Construction—Load and Resistance Factor Design* (Ref. 4). The basic LRFD procedures are essentially the same as those developed by designers of reinforced concrete. The following is a brief summary of the basic elements of the LRFD method.

Loads, Load Factors, and Load Combinations

The various types of loads are as follows:

D = dead load due to the weight of the structure and permanent construction supported by it

L = live load due to occupancy and moveable equipment

L_r = roof live load

W = wind load

S = snow load

E = earthquake load

R = load due to initial rainwater or ice (excludes ponding)

As defined by service conditions, these individual loads constitute the so-called "normal" loading conditions. For design, you first determine appropriate combinations of loads, and then you apply factors to arrive at some increased loading above the service condition, which defines the required strength of the structure. The "design" load is thus qualified as the *factored load combination*. The factored combinations defined by the AISC are

1.4D

1.2D + 1.6L + 0.5 (L_r or S or R)

1.2D + 1.6 (L_r + or S or R) + (0.5L or 0.8W)

1.2D + 1.3W + 0.5L + 0.5 (L_r or S or R)

1.2D + 1.5E + (0.5L or 2.5S)

0.9D − (1.3W or 1.5E)

These are the primary design formulas used to obtain the required design strength for a structure. The specification defines how to increase or decrease these values for special conditions. Keep in mind that code requirements are minimal in nature; for optimal performance, you may need to use higher design strengths. This is a matter of judgment.

Factored Resistance

For design purposes, the required design strength (i.e., the critical factored load combination) is equated to the resistance of the structure. For individual structural elements the resistance is defined in the form of an equation that is expressed by some combination of limiting stress and some properties that relate to the structural element's capacity at failure. Limiting stress for steel is most often its yield strength, although in some situations the ultimate strength may be partially realized. Properties of the element are usually those of the geometric shape of the member's cross section, although for buckling failure the member's unbraced length may also be a consideration.

For example, for a simple tension member its limiting resistance at yield may be expressed as

$$T = (F_y)(A)$$

where F_y = the yield limit of the steel
A = the area of the tension member's cross section

In almost all cases, the member's resistance is also factored (thus the full meaning of LRFD). That is, the resistance is reduced by some amount, expressed in the form of a probability factor, called the *resistance factor* and designated by the Greek letter phi, ϕ. For an actual design procedure, the required strength is equated to the factored resistance and some desired property for the structural member is derived (area, section modulus, moment of inertia, etc.).

The resistance factor varies for different structural elements. Specific procedures for design relate to individual types of structural actions and individual forms of elements.

Serviceability and Other Considerations

Strength is not a designer's only concern. The AISC specification refers to a number of other concerns under the general grouping "Serviceability and Other Considerations." Some specific situations mentioned in the specification are the following:

Impact. Refers to loads that have a true dynamic nature of application; includes wind gusts, seismic movements, and polka dancers. Something more than simple static strength is involved in these situations.

Ponding. Occurs during rainstorms, when water build-up on large flat roof areas is aggravated by sagging of the roof structure (this concern is critical for long-span, very flat and very slow-draining roofs). The ponded water causes more deflection, which in turn allows deeper ponding; before long, a progressive failure can occur.

Deflection. Limiting deflections for service conditions are not related to limiting failure loads; they relate to service load conditions and must be investigated much the same as for the old stress method.

Vibration. Steel structures tend to be light and flexible and tend to transmit energy efficiently. Thus they can be agents for transmission of sounds and vibrations throughout a building. Vibrations also may be intense enough to loosen or fracture connections. These situations need special investigation for various building occupancy conditions.

Drift. Refers to the lateral (horizontal) deflection of multistory frames. These movements may cause structural problems, but more often they affect attached construction, causing cracking of plaster, fracture of glazing, and so on.

Corrosion. The oxygen and moisture in air is sufficient to cause rusting in ordinary steel. Critical structural loss can occur if corrosion is significant. Designers must be aware of potential corrosive conditions—for example, exposure to salt water spray—and understand that protecting the structure is a critical matter.

Fatigue. Usually involves rapidly reversing stress conditions that occur in cycles. If this action extends into steel's plastic range, structures face progressive and accumulative deformations, possibly leading to fracturing.

Fire. As I discuss elsewhere in this book, steel is highly vulnerable when exposed to high heat. As a result, designers must carefully analyze the steel's potential exposure to fire and provide protection as necessary. A minor advantage is steel's lack of combustibility (compared to wood, for example).

The specifications provide extensive requirements for (or at least warn about) these and other potentially critical conditions. Designers must be aware of all these problems, although only a few are significant when designing ordinary building structures. If you follow code requirements for minimum construction and heed ordinary design and construction practices, most of these conditions do not pose any special problems.

5.10 CHOOSING A DESIGN METHOD

As I write this book, you can design using either the ASD or LRFD methods. Used for more than a century, the ASD method has spawned an armada of design aids developed to facilitate rapid design. Most designers today continue to use these aids for preliminary design work and then use the LRFD method for final designs. Exploratory design thrives on quick answers, so anything that helps designers define a trial design is always useful.

In time the LRFD method will become as routine as the strength methods used for concrete structures. Still, understanding simple stress behaviors and stress-strain relationships is easier if you understand simple elastic relationships. Furthermore, designers must examine service load deformations for the lower stress levels—not limiting stress behaviors. The bottom line: Students of structural design must understand ASD and LRFD concepts.

In this book I explain and illustrate both methods. However, I do not provide complete explanations for all situations. I use the ASD method for general presentations because it is simpler; in most cases, I offer some additional illustrations done by LRFD methods for comparison. Most of the computational work in this book is for simple and ordinary structures, so you may see little advantage in using the LRFD method. Thus you should make no judgments about any overall benefits of one method on the basis of the work illustrated here.

6

HORIZONTAL-SPAN FRAMING SYSTEMS

In this chapter I discuss the design of the basic elements found in the most common framing systems used for horizontal-spanning structures for roofs and floors.

6.1 ROLLED SHAPES AS BEAMS

Although several rolled shapes serve beam functions, the most widely used shape is the *wide-flange shape*—that is, that member with an I-shaped cross section, known as a W shape. Except for those members of the W series whose cross section approaches a square (in other words, whose flange width approximately equals its nominal depth), the proportions of W shapes are optimal for flexure about their major axis (designated *x-x*).

When designing beams, designers must heed the following:

Flexural Stress. When designing beams, designers worry most about flexural stresses generated by bending moments. For the W shape, the basic limiting stress used in the ASD method is $0.66F_y$, although some special circumstances may reduce this value. For flexure, the principal beam cross section property is the section modulus, designated S. Resisting moment is expressed as the product of the section modulus and the limiting flexural stress:

$$M = (S)(F_b)$$

If you use the LRFD method, resisting bending moment is based on the fundamental limit F_y, although some special circumstances may change this value. The basic maximum limit for bending moment is the fully plastic moment, for which the significant beam cross section property is the *plastic section modulus*, designated Z. The formula for resisting moment is

$$M = (Z)(F_y)$$

Shear Stress. Although shear stress is critical in wood beams and concrete beams, it is a problem in steel beams only where buckling of the thin beam web is part of a general cross section buckling failure. For beam shear alone, the critical stress is the diagonal compression, which causes buckling in beam action.

Deflection. Although steel is very stiff, steel structures are quite flexible—thus you must investigate the vertical deflection of beams. However, if the span-to-depth ratio of beams falls within certain limits, deflection is rarely critical.

Buckling. In general, inadequately braced beams are subject to various forms of buckling. Especially vulnerable are beams with very thin webs or flanges or beams whose cross sections are especially weak laterally (on the minor axis, or designated *y-y*). Although designers can use reduced stress (ASD) or reduction factors (LRFD) to limit buckling, the most effective solution is to provide adequate bracing.

Connections and Supports. Framed structures have many joints. To ensure the transfer of necessary structural forces through the joints, designers must pay attention to the details of the connections and supports.

When designing a beam, you must consider not only a beam's basic functions, but also how it plays a part in overall system actions and interactions. In this chapter I focus on individual beam actions; in other chapters I describe how beams affect structural systems.

Selecting the optimal shape for a given situation is never a cut-and-dried task: current AISC manuals list several hundred W shapes (of course, several other shapes serve beam functions in special circumstances). Other things being equal, designers often choose the most economical shape—usually the one that weighs the least because steel is generally priced by unit weight.

Note: Just as a beam sometimes must develop other actions—such as tension, compression, or torsion—other structural elements may develop beam actions. For example, walls may span for bending against wind pressures; columns may receive bending moments as well as compression loads; and truss chords may span as beams as well as function for basic truss actions. In this chapter I desribe basic beam *functions*.

6.2 BENDING OF BEAMS

When designing a beam for bending, designers usually must determine the maximum bending moment. Because formulas for resisting moment rely on properties of the member's cross section, designers often use such formulas to determine desirable cross sections.

Allowable Stress Design

Use the basic flexural formula ($f = M/S$) to determine the minimum section modulus required. Because weight depends on area, not section modulus, a beam may have more section modulus than required and still be the most economical choice. The following example illustrates the basic procedure.

Example 1. Design a simply supported beam to carry a superimposed load of 2 kips per ft [29.2 kN/m] over a span of 24 ft [7.3 m]. (The term *superimposed load* denotes any load other than the weight of a structural member.) The allowable bending stress is 24 ksi [165 MPa].

Solution: The bending moment caused by the superimposed load is

$$M = \frac{wL^2}{8} = \frac{2 \times (24)^2}{8} = 144 \text{ kip-ft} \quad [195 \text{ kN-m}]$$

The required section modulus for this moment is

$$S = \frac{M}{F_b} = \frac{144 \times 12}{24} = 72.0 \text{ in.}^2 \quad [1182 \times 10^3 \text{ mm}^3]$$

Table A.1 lists a W 16 × 45 with a section modulus of 72.7 in.³ [1192 × 10³ mm³], but this section modulus is so close to that required that it provides almost no margin for the effect of the beam weight. The table also includes a W 16 × 50 with an *S* of 81.0 in.³ [1328 × 10³ mm³] and a W 18 × 46 with an *S* of 78.8 in.³ [1291 × 10³ mm³]. In the absence of any known restriction on beam depth, try the lighter section. The bending moment at the center of the span with this beam is

$$M = \frac{wL^2}{8} = \frac{46 \times (24)^2}{8}$$

$$= 3312 \text{ ft-lb or } 3.3 \text{ kip-ft} \quad [4.46 \text{ kN-m}]$$

Thus the total bending moment at midspan is

$$M = 144 + 3.3 = 147.3 \text{ kip-ft} \quad [199.5 \text{ kN-m}]$$

The section modulus required for this moment is

$$S = \frac{M}{F_b} = \frac{147.3 \times 12}{24} = 73.7 \text{ in.}^3 \quad [1209 \times 10^3 \text{ mm}^3]$$

Because this required value is less than that of the W 18 × 46, this section is acceptable.

To select a rolled shape on the basis of required section modulus, you can use AISC tables (Ref. 3) that list beam shapes in descending order of section modulus value.

Table B.1 contains data from such AISC tables. The shapes whose designations are listed in bold have an especially cost-efficient bending moment resistance; other sections have the same or smaller section modulus but are heavier and thus more expensive.

Table B.1 also includes data vital to considering a beam's lateral support: the two limiting lengths L_c and L_u. If you assume the maximum allowable stress of 24 ksi [165 MPa], the required section modulus is proper only for beams whose lateral unsupported length is equal to or less than L_c.

Table B.1 also lists values for the maximum bending resistance (M_R) of the sections of A36 steel beams. Although the condition of the noncompact section may be noted, in this case the M_R values already take into account the reduced values for bending stress.

Example 2. Rework Example 1 by using Table B.1.

Solution: As before, determine that the bending moment due to the superimposed loading is 144 kip-ft [195 kN-m]. Because some additional M_R capacity is required to account for the beam's weight, scan the table for shapes with an M_R of slightly more than 144 kip-ft [195 kN-m]:

Shape	M_R (kip-ft)	M_R (kN-m)
W 21 × 44	162	220
W 16 × 50	160	217
W 18 × 46	156	212
W 12 × 58	154	209
W 14 × 53	154	209

The W 21 × 44 is the lightest section, but you must take into account all other design considerations, such as restricted depth, before choosing.

Note: Table B.1 does not include all the W shapes listed in Table A.1; excluded are the shapes that are almost square and thus used ordinarily for columns rather than beams.

For the following problems, design for bending stress only. Use A36 steel with an allowable bending stress of 24 ksi [165 MPa] and choose the lightest member for each case.

Problem 6.2.A. Design a simple beam 14 ft. [4.3 m] long that has a total uniformly distributed load of 19.8 kips [88 kN].

Problem 6.2.B. Design a beam with a span of 16 ft [4.9 m] that has a concentrated load of 12.4 kips [55 kN] at the span's center.

Problem 6.2.C. A beam 15 ft [4.6 m] long has three concentrated loads of 4 kips, 5 kips, and 6 kips at 4 ft, 10 ft, and 12 ft [17.8 kN, 22.2 kN, and 26.7 kN at 1.2 m, 3 m, and 3.6 m], respectively, from the left-hand support. Design the beam.

Problem 6.2.D. A beam 30 ft [9 m] long has concentrated loads of 9 kips [40 kN] each at the third points and also a total uniformly distributed load of 30 kips [133 kN]. Design the beam.

Problem 6.2.E. Design a beam 12 ft [3.6 m] long that has a uniformly distributed load of 2 kips/ft [29 kN/m] and a concentrated load of 8.4 kips [37.4 kN] 5 ft [1.5 m] from a support.

Problem 6.2.F. A beam 19 ft [5.8 m] long has concentrated loads of 6 kips [26.7 kN] and 9 kips [40 kN] at 5 ft [1.5 m] and 13 ft [4 m], respectively, from the left-hand support. In addition, it has a uniformly distributed load of 1.2 kips/ft [17.5 kN/m] beginning 5 ft [1.5 m] from the left support and continuing to the right support. Design the beam.

Problem 6.2.G. A steel beam 16 ft [4.9 m] long has two uniformly distributed loads, one of 200 lb/ft [2.92 kN/m] extending 10 ft [3 m] from the left support and one of 100 lb/ft [1.46 kN/m] extending over the remainder of the beam. In addition, the beam has a concentrated load of 8 kips [35.6 kN] at 10 ft [3 m] from the left support. Design the beam.

Problem 6.2.H. Design a simple beam 12 ft [3.7 m] long that has two concentrated loads of 12 kips [53.4 kN] each, one 4 ft [1.2 m] from the left end and the other 4 ft [1.2 m] from the right end.

Problem 6.2.I. A cantilever beam 8 ft [2.4 m] long has a uniformly distributed load of 1600 lb/ft [23.3 kN//m]. Design the beam.

Problem 6.2.J. A cantilever beam 6 ft [1.8 m] long has a concentrated load of 12.3 kips [54.7 kN] at its unsupported end. Design the beam.

Load and Resistance Factor Design

Establish the limiting moment resistance by developing the full plastic stress range for the beam cross section. *Note:* A beam's actual load limit is not determined strictly by flexure; other failures may occur before the flexural limit is reached. For more information, refer to the following sections; here I limit the discussion to flexural resistance.

The maximum resisting moment predicted by elastic theory occurs when the stress at the extreme fiber reaches the elastic yield value, F_y, expressed as

$$M_y = F_y S$$

Beyond this condition an inelastic, or *plastic*, stress condition develops on the beam cross section, so you cannot express the resisting moment by elastic theory equations.

Figure 6.1 represents an idealized form of a load-test response for a ductile steel specimen. Up to the yield point, the deformations are proportional to the applied stress; beyond the yield point, the steel experiences a deformation without an increase in stress. For A36 steel this additional deformation (called the *plastic range*) is approximately *15* times that produced just before yield occurs, which is why A36 steel is considered ductile.

Beyond the plastic range the material again stiffens (known as the *strain hardening* effect), losing its ductility. From that point to the *ultimate stress* limit is a second range in which additional deformation is produced only by increased stress.

For plastic failure to be significant, the plastic range must be several times that of the elastic range. The plastic theory generally applies only to steels with a yield point not exceeding 65 ksi [450 MPa]—as the yield limit increases, the plastic range decreases.

The following example illustrates how designers use the elastic theory.

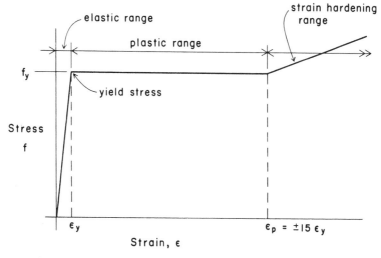

FIGURE 6.1 Idealized form of ductile steel's stress-strain response.

Example 3. A simple beam has a span of 16 ft [4.88 m] and supports a single concentrated load of 18 kips [80 kN] at its center. If the beam is a W 12 × 30 and is adequately braced to prevent buckling, compute the maximum flexural stress.

Solution: Determine the maximum value of the bending moment

$$M = \frac{PL}{4} = \frac{18 \times 16}{4} = 72 \text{ kip-ft} \quad [98 \text{ kN-m}]$$

From Table A.2, S is 38.6 in.^{3} [632 × 10^3 mm^3]. Thus the maximum stress is

$$f = \frac{M}{S} = \frac{72 \times 12}{38.6} = 22.4 \text{ ksi} \quad [154 \text{ MPa}]$$

Figure 6.2*d* shows that this stress condition occurs only at the beam section at midspan. Figure 6.2*e* shows the form of the deformations that accompany the stress condition. This stress level is well below the elastic stress limit (i.e., yield point) and, in this example, below the allowable stress of 24 ksi.

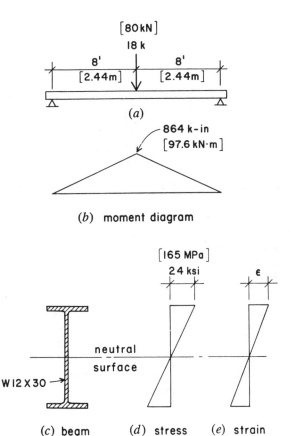

(a)

(b) moment diagram

(c) beam   (d) stress   (e) strain

FIGURE 6.2 Beam's elastic behavior.

Figure 6.3a shows the limiting moment expressed in allowable stress terms, which occurs when the maximum flexural stress reaches the yield stress limit.

If the loading that causes the yield limit flexural stress is increased, a stress condition (see Figure 6.3b) develops as the ductile material deforms plastically. When the higher stress level spreads over the beam cross section, a resisting moment in excess of M_y has has developed. Given a high level of ductility, a limit for this condition takes the form shown in Fig. 6.3c, and the limiting resisting moment is described as the *plastic moment*, designated M_p.

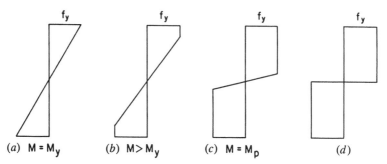

FIGURE 6.3 Flexural stress from elastic range to plastic range.

Although a small percentage of the cross section near the beam's neutral axis remains in an elastic stress condition, its effect on the development of the resisting moment is negligible. Thus you may assume that the full plastic limit is developed by the condition shown in Figure 6.3*d*.

Attempts to increase the bending moment beyond M_p cause large rotational deformation—the beam acts as though it were hinged. For practical purposes, therefore, consider the resisting moment capacity of the ductile beam to be exhausted at the plastic moment; additional loading merely causes a free rotation at the location of the plastic moment, known as the *plastic hinge* (see Figure 6.4).

The resisting plastic moment is expressed as

$$M = F_y \times Z$$

where Z is the *plastic section modulus*. Calculating Z is a convoluted exercise:

1. In Figure 6.5, which shows a W shape subjected to a flexural stress corresponding to the fully plastic section,

   A_u = the upper area of the cross section, above the neutral axis

   y_u = distance of the centroid of A_u from the neutral axis

   A_l = the lower area of the cross section, below the neutral axis

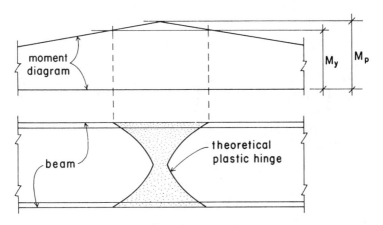

FIGURE 6.4 Plastic hinge.

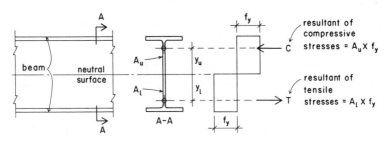

FIGURE 6.5 Plastic resisting moment.

y_l = distance of the centroid of A_l from the neutral axis

You can express the equilibrium of the internal forces on the cross section (the resulting forces C and T developed by the flexural stresses) as

$$\Sigma H = 0$$

or

$$[A_u \times (+f_y)] + (A_l \times (-f_y)] = 0$$

so

$$A_u = A_l$$

Thus the plastic stress neutral axis divides the cross section into equal areas, whether the section is symmetrical or unsymmetrical.

2. The resisting moment equals the sum of the moments of the stresses. You can express M_p as

$$M_p = (A_u \times f_y \times y_u) + (A_l \times f_y \times y_l)$$

or

$$M_p = f_y [(A_u \times y_u) + (A_l \times y_l)]$$

or

$$M_p = f_y \times Z$$

The quantity $[(A_u \times y_u) + (A_l \times y_l)]$ is the property of the cross section defined as the plastic section modulus, designated Z.

Using the expression for Z just derived, you can compute Z for any cross section. In this book, however, you only need to find Z for W shapes and Table A.1 lists all such values, along with the values of properties used for elastic analysis.

For the same W shape Z_x is larger than S_x. In the following example I show why, comparing the fully plastic resisting moment to the yield stress limiting moment by elastic stress.

Example 4. A simple beam consisting of a W 21 × 57 is subjected to bending. Find the limiting moments (a) based on elastic stress conditions and a limiting stress of $F_y = 36$ ksi (b) based on full development of the plastic moment.

Solution: For (a) the limiting moment is

$$M_y = F_y \times S_x$$

From Table A.1, the value of S_x for the W 21 × 57 is 111 in.[3], so the limiting moment is

$$M_y = (36) \times (111) = 3996 \text{ kip-in. or } \frac{3996}{12} = 333 \text{ kip-ft}$$

From Table A.1, $Z_x = 129$ in.[3], so for (b) the limiting plastic moment is

$$M_p = F_y \times Z = (36)(129) = 4644 \text{ kip-in.}$$

$$\text{or } \frac{4644}{12} = 387 \text{ kip-ft}$$

The increase in moment resistance is $387 - 333 = 54$ kip-ft, or a gain of $(54/333)(100) = 16.2\%$.

Although Z_x is greater than S_x, it's not easy to demonstrate why it's better to use the plastic moment for design. You must use a different process regarding safety factors—and if you use the LRFD method, a whole different approach. Still, you will find significant differences when designing continuous beams, restrained beams, and rigid column-beam frames.

Figure 6.6 shows a uniformly distributed load of w lb per ft on a beam that is fixed at both ends. The moment induced by this condition is distributed along the beam length in the manner of a symmetrical parabola with maximum height (maximum moment) of $w L^2/8$ (see Figure B.2, Case 2). For other conditions of support or continuity, the distribution pattern changes but the total moment remains the same.

As shown in Figure 6.6a, fixed ends cause the distribution shown beneath the beam, with maximum end moments of $w L^2/12$ and a moment at the center of $w L^2/8 - w L^2/12 = w L^2/24$. This distribution continues as long as stress conditions do not exceed the beam's yield limit. As a result, the limiting condition for elastic conditions is when a load limit of w_y corresponds to the yield stress limit (see Figure 6.6b).

Once the flexural stress at the point of maximum moment reaches the fully plastic state, further loading results in a plastic hinge, so the resisting moment there will never exceed the plastic moment. With additional loading, the moment at the plastic hinge

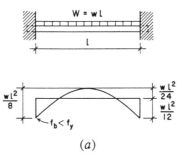

(a)

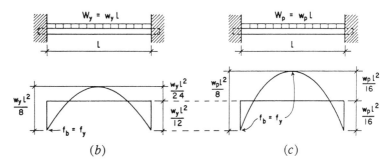

(b)   (c)

FIGURE 6.6 Fully plastic beam.

remains constant, but another plastic hinge may develop at another location.

Figure 6.6c shows the plastic limit for a beam fixed at both ends; this condition occurs when both maximum moments are equal to the beam's plastic limit. Thus, if $2(M_p) = w_p L^2/8$, then the plastic limit (M_p) is equal to $w_p L^2/16$. The following example illustrates an LRFD investigation.

Example 5. A beam with fixed ends carries a uniformly distributed load. The beam consists of a W 21 × 57 of A36 steel with $F_y = 36$ ksi. Find the uniform load if (a) the flexure limit is the beam's elastic limit (b) the beam develops the fully plastic moment at critical moment locations.

Solution: As determined in Example 4,

M_y = 333 kip-ft (elastic stress limit at yield)
M_p = 387 kip-ft (fully plastic moment)

(a) From Figure 6.6b, the maximum moment for elastic stress is w $L^2/12$. If this limit equals the limiting value for moment,

$$M_y = 333 = \frac{w_y L^2}{12}$$

from which

$$w_y = \frac{(333)(12)}{L^2} = 3996/L^2 \quad \text{(in kip-in. units)}$$

(b) From Figure 6.6c, the maximum moment for the plastic moments with plastic hinging at the fixed ends is $w L^2/16$. If this limit equals the limiting value for moment,

$$M_p = 387 = \frac{w_p L^2}{16}$$

from which

$$w_p = \frac{(387)(16)}{L^2} = 6192/L^2 \quad \text{(in kip-in. units)}$$

If you combine the increase due to the plastic moment with the effect of the redistribution of moments due to plastic hinging, the total increase is $6192 - 3996 = 2196/L^2$, and the percentage gain is

$$\frac{2196}{3996}(100) = 55\%$$

Note: This gain is substantially greater than that calculated in Example 4, where I considered only the difference in moments. It is this combined effect that is significant when applying plastic analysis and the LRFD method for continuous structures.

Problem 6.2.K. A simple-spanning, uniformly loaded beam (Table B.2, Case 2) consists of a W 18 × 50 with F_y = 36 ksi. Find the percentage gain in the limiting bending moment if a fully plastic condition is assumed instead of a condition limited by elastic stress.

Problem 6.2.L. If the beam in Problem 6.6.K has fixed ends instead of simple supports, find the percentage gain in load carrying capacity if a fully plastic condition is assumed rather than a condition limited by elastic stress.

Note: I further discuss the plastic moment in Chapters 7 and 8.

6.3 SHEAR IN BEAMS

Shear in beams is the vertical slicing effect produced by the opposition of vertical loads on the beams (down) and the reactive forces at the beam supports (up). To illustrate the internal shear force mechanism, designers use a shear diagram. When a simply supported beam has a uniformly distributed load, such a diagram looks like Figure 6.7a.

This load condition causes an internal shear force that peaks at the beam supports and steadily decreases to zero at the center of the beam span. For a beam with a constant cross section throughout the span, the critical location for shear is at the supports; in fact, everywhere else shear is not a concern. *Note:* Because this loading condition is common for many beams, you need investigate only the support conditions for such beams.

Figure 6.7b shows another common condition: a major concentrated load within the beam span generates a major internal shear force over some length of the beam. If the concentrated load is close to one support, a critical internal shear force is created between the load and that support. An example of this condition: Framing arrangements for roof and floor systems frequently employ beams that carry the end reactions of other beams.

Internal shear force develops shear stresses in a beam. How these stresses are distributed over the cross section depends on the beam cross section—its general form and its geometric properties. For a simple rectangular cross section, such as with a wood beam,

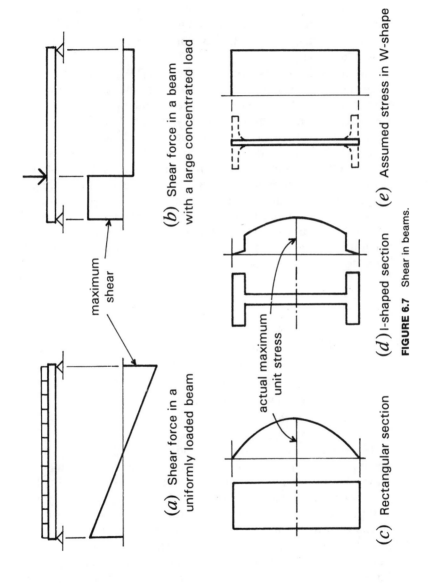

(a) Shear force in a uniformly loaded beam

(b) Shear force in a beam with a large concentrated load

(c) Rectangular section

(d) I-shaped section

(e) Assumed stress in W-shape

maximum shear

actual maximum unit stress

FIGURE 6.7 Shear in beams.

the distribution of beam shear stress is parabolic (see Figure 6.7c), with a maximum shear stress at the beam neutral axis and zero stresses at the extreme fiber distances (i.e., top and bottom edges). This shear stress is critical to designers of solid timber beams because it can split the beam horizontally near the neutral axis, along the wood grain.

For the I shaped cross section of the typical W shape rolled steel beam, the beam shear stress distribution is as shown in Figure 6.7d (the "derby hat" form). As with the rectangular cross section, the shear stress peaks at the beam neutral axis, but the falloff is less rapid between the neutral axis and the inside of the beam flanges. Although the flanges take some shear force, the sudden increase in beam width results in an abrupt drop in the beam unit shear stress. As a result, when designers investigate shear stress for the W shape, they traditionally ignore the flanges and assume the beam's shear-resisting portion to be an equivalent vertical plate (see Figure 6.7e) with a width equal to the beam web thickness and a height equal to the full beam depth. In turn, designers can establish an allowable value for a unit shear stress and compute the actual stress as

$$ f_v = \frac{V}{t_w \, d_b} $$

where f_v = average unit shear stress, based on an assumed distribution (as in Figure 6.7e)

V = internal shear force at the cross section

t_w = beam web thickness

d_b = overall beam depth

Uniformly loaded beams are seldom critical with regard to shear stress on the basis just described. The most common beam support is shown in Figure 6.8a: a connecting device effects the transfer of the end shear force to the beam support (commonly using a pair of steel angles that grasp the beam web and are turned outward to fit flat against another beam's web or the side of a column). Welding a connecting device to the supported beam's web actually reinforces the web at this location; thus the critical shear stress is at that portion of the beam web just beyond the connector.

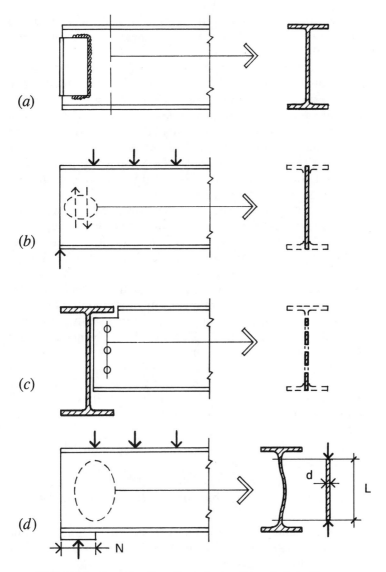

FIGURE 6.8 Considerations for end support in rolled steel beams.

The shear force is as shown in Figure 6.8*b*, and it is assumed to operate on the effective section of the beam as discussed previously.

However, other situations can cause critical conditions for the whole transfer of vertical force at the end of the supported beam. When the supporting element is also a W shape, and the tops of the beams are at the same level (common in framing systems), you must cut back the top flange and part of the supported beam's web to permit the web's end to get as close as possible to the side of the supporting beam's web (Figure 6.8*c*). Doing so results in some loss of the full shear-resisting area assumed in Figure 6.8*b* and an increase in the unit shear stress.

You also can reduce the shear-resisting area by using bolts rather than welds to fasten connecting angles to the supported beam's web (see Figure 6.8*c*). The full reduction of the shear-resisting area will thus include the losses due to the bolt holes and the notched top of the beam.

However, shear stress may not be critical at a beam support. Figure 6.8*d* shows a different means of support, consisting of a bearing of the beam end on top of the support, usually in this case the top of a wall or a wall ledge. The potential problem has more to do with vertical compression force, which squeezes the beam end and causes a column-like action in the thin beam web. In fact, this force actually may produce a column-like form of failure; Figure 6.9, which compares the compression capacity to the web's relative slenderness, illustrates three such cases:

- A very stiff (thick) web, which may almost fulfill the material's full yield stress limit
- A somewhat slender web, which fails from some combined yield stress and buckling effect (called an *inelastic buckling* response)
- A very slender web, which fails in the classic Euler formula manner (called *elastic buckling*); basically a deflection failure rather than a stress failure

Although the web cross section is a major element in each response, another factor affects the column-like responses: how much of the web length—or the beam length—along the span is involved in the response to the vertical force effect. In Figure 6.8*d*, this factor is represented by *N*, the length dimension of the bearing plate along the beam length. Now you can identify how much of the total beam web (in three dimensions) is significantly involved in the column action.

Given all the preceding investigations, you may want to choose a beam shape whose web is sufficient. However, other criteria (flexure, deflection, framing details, and so on) may indicate that you should choose a vulnerable web. If so, you may need to *reinforce* the web. (The usual means is to insert vertical plates on either side of the web and then fasten them to the web as well as to the beam flanges. These plates then brace the slender web (column) and absorb some of the vertical compression stress in the beam.)

I further discuss the general problem of vertical force effects on beam webs in Section 6.9. I cover various problems regarding framing connections in Chapter 11.

For practical design purposes, you can handle the beam end shear and end support limits of unreduced webs with data from AISC tables, which I demonstrate how to use in Section 6.9.

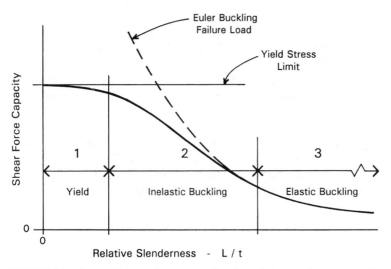

FIGURE 6.9 Compression resistance as related to relative slenderness; transition from stress to stiffness limits.

Note: All the phenomena and problems mentioned in this section are treated similarly in both the ASD and LRFD methods; the only difference is whether you establish safety by adjusting stresses or loads.

6.4 DEFLECTION OF BEAMS

Designers hope to control structural deformations; they worry especially about whole structural elements or assemblages.

With a modulus of elasticity of 29,000 ksi, steel is 8 to 10 times as stiff as average structural concrete and 15 to 20 times as stiff as structural lumber, but steel structures are frequently deformable and flexible. The reasons for this apparent contradiction are simple: Because steel is expensive, designers usually form it into elements with thin parts (beam flanges and webs, for example), and because steel is very strong, designers frequently form it into relatively slender elements (e.g., beams and columns).

For a beam in a horizontal position, the critical deformation is usually the maximum sag, called the beam's *deflection*. For most beams this deflection is too small to see. However, any load on a beam, including the beam's own weight, causes some deflection (see Figure 6.10). In the case of the simply supported, symmetrical, single-span beam in Figure 6.4, the maximum deflection occurs at midspan; although this deformation is usually the only value of concern for design, as the beam deflects, its ends rotate (unless restrained) and this deformation may also be of concern in some situations.

When deflection is excessive, designers usually select a deeper beam. The critical property of the beam cross section is its *moment of inertia* (I) about its major axis (I_x for a W shape), which is typically affected significantly by increases in the beam's depth. Formulas for beam deflection take the following form:

$$\Delta = C \frac{W L^3}{E I}$$

where Δ = the deflection, measured vertically in in. or mm.

C = a constant related to the form of the load and support conditions for the beam

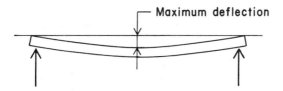

FIGURE 6.10 Deflection of a simple beam under symmetrical loading.

W = the load on the beam

L = the span of the beam

E = the modulus of elasticity of the material of the beam

I = the moment of inertia of the beam cross section for the axis about which bending occurs

Note that the deflection is directly proportional to the load. In addition, the deflection is proportional to the third power of the span—that is, double the span and you get 2^3 or eight times the deflection. Increases in either the material's stiffness or the beams geometric form (I) cause direct proportional reduction of the deflection. Since E is constant for all steel, designers must deal only with the beam's shape to modulate deflections.

Excessive deflection causes various problems. For roofs, an excessive sag may disrupt the intended drainage patterns. For floors, deflection can cause bounce. For a simple-span beam (see Figure 6.10), the usual concern is for the maximum sag at the beam midspan. For a beam with a projected (cantilevered) end, deflection may create a problem at the unsupported cantilevered end— whether the problem involves downward deflection (see Figure 6.11a) or upward deflection (Figure 6.11b) depends on the extent of the cantilever. For continuous beams, load in any span causes some deflection in all spans; this problem is especially critical when loads vary in different spans or the span lengths differ significantly (see Figure 6.11c).

Most deflection problems in buildings crop up when a member's structural deformation affects adjacent or supported elements. For example, when beams are supported by other beams (known as *girders*), excessive rotation due to deflection at the ends of the supported beams can cause cracking in the floor deck that is

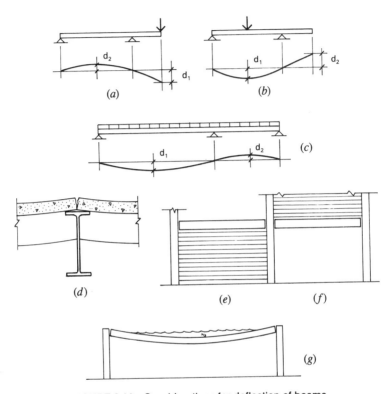

FIGURE 6.11   Considerations for deflection of beams.

continuous over the girders (see Figure 6.11d). In addition, an accumulative deflection—due to independent deflections of the deck, beams, and girders—can affect floors and roofs.

An especially difficult problem is the effect of beam deflections on nonstructural construction elements. In Figure 6.10e, for example, a beam lies directly over a solid wall; if the wall fits tightly beneath the beam, any beam deflection causes the beam to bear on top of the wall, which is unacceptable if the wall is relatively fragile (a metal and glass curtain wall, for example). Another example: Relatively rigid walls (plastered walls, for example) supported by spanning beams (see Figure 6.11f), are intolerant of *any* deformation, so any significant sag of the beam is really critical.

For long-span structures (an ambiguous class, usually meaning 100 ft or more span), a flat roof surface is especially vulnerable. Even if designers adhere to code-mandated minimum drainage requirements, heavy rain may linger on a roof. This unexpected loading can cause deflection of the spanning structure; in turn, the sag may create a depressed area that collects the rain, forming a pond (see Figure 6.11g), which constitutes an additional load, causing more sag—and a deeper pond. Because this cycle can cause a structure to fail, codes (including the AISC Specification) now require designers to investigate potential ponding.

Allowable Deflections

What is permissible for beam deflection—to avoid the problems illustrated in Figure 6.11—is mostly a matter of judgment. After investigating each situation, designers must cooperate and collectively decide which design controls are necessary.

For spanning beams in ordinary situations, designers have developed some rules of thumb, expressed as maximum allowable beam curvature. These rules, described as a limiting ratio of the deflection to the beam span (L), are expressed as a fraction of the span. Some typical limits are

1/150:	Total load deflection to avoid visible sag on short to medium spans
1/180:	Total load deflection of a roof structure
1/240:	Deflection under live load only for a roof structure
1/240:	Total load deflection of a floor structure
1/360:	Deflection under live load only for a floor structure

Deflection of Uniformly Loaded Simple Beams

The beam used most often in flat roof and floor systems is the uniformly loaded beam with a single, simple span (i.e., no end restraint), as in Figure B.2, Case 2. You can quantify the beam's behavior as follows:

Maximum bending moment is

$$M = \frac{WL}{8}$$

Maximum stress on the beam cross section is

$$f = \frac{Mc}{I}$$

Maximum midspan deflection is

$$\Delta = \frac{5}{384} \times \frac{WL^3}{EI}$$

From these relationships, you can derive a convenient formula for deflection.

Given that the dimension c in the bending stress formula is $d/2$ for symmetrical shapes, you know that

$$f = \frac{Mc}{I} = \left(\frac{WL}{8}\right)\left(\frac{d/2}{I}\right) = \frac{WLd}{16I}$$

Thus

$$\Delta = \frac{5\,WL^3}{384\,EI} = \left(\frac{WLd}{16\,I}\right)\left(\frac{5\,L^2}{24\,Ed}\right) = (f)\left(\frac{5\,L^2}{24\,Ed}\right) = \frac{5fL^2}{24\,Ed}$$

This formula works for any beam symmetrical about its bending axis.

Next, substitute 24 ksi for f (a common limit for bending stress for W shapes) and 29,000 ksi for E (steel's modulus of elasticity). Also, for convenience, add a factor of 12 because spans are usually measured in feet, not inches. Thus

$$\Delta = \frac{5fL^2}{24\,Ed} = \left(\frac{5}{24}\right)\left(\frac{24}{29{,}000}\right)\left(\frac{[12\,L]^2}{d}\right) = \frac{0.02483\,L^2}{d}$$

In metric units, $f = 165\,\text{MPa}$ and $E = 200\,\text{GPa}$. Also, convert the span to meters:

$$\Delta = \frac{0.0001719\,L^2}{d}$$

Examples

The following examples illustrate how to investigate the uniformly loaded simple beam for deflection.

Example 1. A simple beam has a span of 20 ft [6.10 m] and a total uniformly distributed load of 39 kips [173.5 kN]. The beam is a steel W 14 × 34. Find the maximum deflection.

Solution: First determine the maximum bending moment as

$$M = \frac{WL}{8} = \frac{(39)\,(20)}{8} = 97.5 \text{ kip-ft}$$

From Table A.1, $S = 48.6$ in.^{3}, so the maximum bending stress is

$$f = \frac{M}{S} = \frac{(97.5)\,(12)}{48.6} = 24.07 \text{ ksi}$$

Because the limiting stress is 24 ksi, you may consider the beam stressed to its limit. Thus you may use the derived formula. From Table A.1, the beam's true depth is 13.98 in. Then

$$\Delta = \frac{0.02483\,L^2}{d} = \frac{(0.02483)\,(20)^2}{13.98} = 0.7104 \text{ in.} \quad [18.05 \text{ mm}]$$

To double check your work, use the general formula for deflection of the simple beam with uniformly distributed load. Since $I = 340$ in.^{4},

$$\Delta = \frac{5\,WL^3}{EI} = \frac{5(39)\,(20 \times 12)^3}{(29{,}000)\,(340)} = 0.712 \text{ in.}$$

which is close enough.

Rarely is the chosen beam stressed precisely at 24 ksi. The following example illustrates a more typical situation.

Example 2. A simple beam consisting of a W 12 × 26 carries a total uniformly distributed load of 24 kips [107 kN] on a span of 19 ft [5.79 m]. Find the maximum deflection.

Solution: Find the maximum bending moment:

$$M = \frac{WL}{8} = \frac{(24)(19)}{8} = 57 \text{ kip-ft}$$

From Table A.1, S for the beam is 33.4 in.^{3}, so the maximum bending stress is

$$f = \frac{M}{S} = \frac{(57)(12)}{8} = 20.48 \text{ ksi}$$

When using the deflection formula based only on span and beam depth, the basis for bending stress is 24 ksi. Therefore you must make an adjustment: the ratio of true bending stress to 24 ksi. Thus

$$\Delta = \frac{20.48}{24} \times \frac{0.02483\, L^2}{d}$$

$$= 0.8533 \times \frac{0.02483\,(19)^2}{12.22} = 0.626 \text{ in. } [16 \text{ mm}]$$

You can use the derived deflection formula involving only span and beam depth to plot a graph that displays the deflection of a beam of a constant depth for a variety of spans. Figure 6.12 is a series of such graphs, for beams with depths 6 to 36 in. You can use such graphs to determine beam deflections. In fact, I urge you to verify that deflections found from the graphs for the beams in Examples 1 and 2 reasonably match (i.e., ±5 percent) the computed results. You also can use such graphs in the design process.

For example, once you know the span, you may determine from the graphs what beam depth is required for a given deflection. Often the limiting deflection is given as a limiting percentage of the span (1/240, 1/360, etc.); note how the graph includes lines that represent the usual percentage limits of 1/360, 1/240, and 1/180.

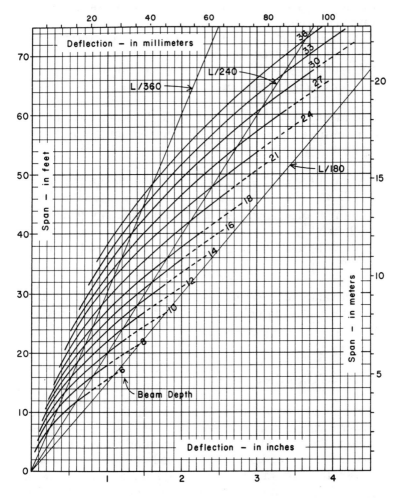

FIGURE 6.12 Deflection of steel beams with a bending stress of 36 ksi [165 MPa].

Say that you need a beam for a 36 ft span, and assume that the total load deflection limit is $L/240$: note how the lines for a span of 36 ft and a ratio of 1/240 intersect near the curve representing an 18 in. deep beam. You know, therefore, that an 18 in. deep beam deflects almost precisely 1/240th of the span if stressed in bending to 24 ksi. The bottom line: Any beam with less depth is inadequate for deflection, and any beam with more depth is conservative in regard to deflection.

Determining deflections for other beams, however, is much more complicated. Nonetheless, many handbooks (including the AISC Manual) provide formulas designers can use to compute deflections for a variety of beam loading and support situations.

Problems 6.4.A,B,C. Find the maximum deflection in inches for the following simple beams of A36 steel with uniformly distributed load;

   (A) W 10 × 33, span = 18 ft, total load = 30 kips [5.5 m, 133 kN]

   (B) W 16 × 36, span = 20 ft, total load = 50 kips [6 m, 222 kN]

   (C) W 18 × 46, span = 24 ft, total load = 55 kips [7.3 m, 245 kN]

Find the values, using (a) the equation for Case 2 in Figure B.2; (b) the formula involving only span and beam depth; (c) the curves in Figure 6.12.

6.5 BUCKLING OF BEAMS

Buckling is a problem for beams that are relatively weak on their transverse axis—that is, the axis of the beam cross section perpendicular to the axis of bending. This condition rarely occurs with concrete beams, but is common with wood and steel beams, as well as trusses that perform beam functions. The cross sections shown in Figure 6.13 belong to members that are susceptible to buckling in beam action.

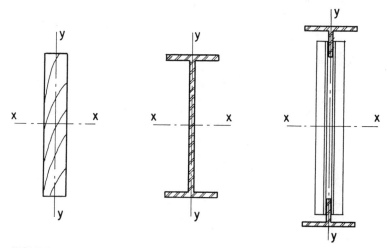

FIGURE 6.13   Beam shapes with low resistance to lateral bending and buckling.

Whether the steel W shape risks buckling depends on the proportions of its dimensions. When the beam flange width (designated b_f in Table A.1) is less than one-half the beam's depth (designated d in Table A.1), buckling resistance becomes critical. Another indicator is the ratio of the section modulus S_x for the major axis to S_y for the minor axis; in general, the lightest shapes in each nominal depth group have the least lateral strength.

When bending resistance may be limited by buckling, the preferred solution is to brace the member, preventing a buckling response to the load. In fact, bracing counteracts the three major buckling forms that can occur with a simple beam.

Figure 6.14*a* shows the response described as *lateral buckling*, which is caused by a compression column-like action of the top portion of the beam. As with any slender, linear, compression element, a sideways buckling is the usual form of failure. For a completely unbraced beam, the midspan is the most critical location for this buckling form; as a result the logical solution is to provide lateral bracing perpendicular to the beam and at the top of the beam. However, you may need to brace a really skinny beam more than once along its length.

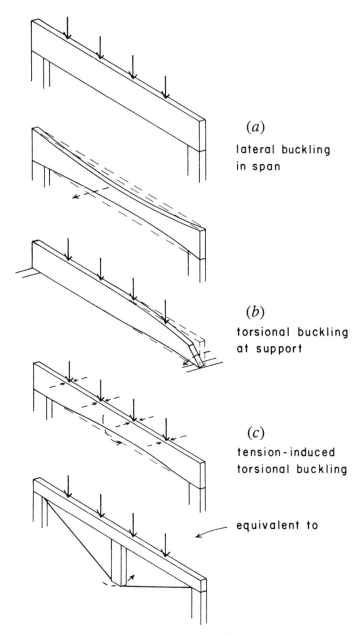

(a)

lateral buckling
in span

(b)

torsional buckling
at support

(c)

tension - induced
torsional buckling

equivalent to

FIGURE 6.14 Buckling of beams: (a) lateral (b) torsion at end (c) torsion in span.

To brace the beam against lateral buckling is to prevent its sideways movement. As with a column, you need very little force to effect this bracing—typically less than 3 percent of the compression force in the member. Indeed, often other construction elements provide this bracing adequately, so no extra bracing is required.

Torsional buckling typically takes two forms. The first occurs at the beam supports (see Figure 6.14*b*), where the beam may roll over in a twisting (i.e., torsional) failure. To brace for this effect is to prevent lateral movement in general for the full beam depth. Again, other construction elements—including attached decking and the beam end connections—often provide this bracing sufficiently.

The other form of torsional buckling occurs within the beam span, a result of tension in the beam's bottom portion. For example, unless the simple truss pictured at the bottom of Figure 6.14 remains in perfect vertical alignment with the loads, failure is highly likely—in the form of a sudden sideways movement by the strut. Similarly, to prevent a twisting, torsional failure of a laterally unbraced beam, you must brace both the bottom and top of the beam.

Design of Laterally Unsupported Beams

The AISC Specification describes how to deal with beam buckling. As with other buckling forms, it prescribes a three-stage progression, in which some effect is tolerated without reduction (in this case, loss of moment resistance):

1. For a limited length, defined as L_c, no reduction in moment capacity is required.
2. When the unbraced length (that is, the length along the beam for which no lateral bracing is provided) exceeds L_c, a small percentage reduction is required, which is sufficient until the unbraced length reaches a second limit, defined as L_u.
3. For lengths exceeding L_u, the AISC Specification gives a formula for reductions that is an adaptation of the Euler buckling load formula for elastic buckling.

The AISC Manual (Ref. 3) includes charts that plot the various limits for lateral unsupported length against yield values for reduced moment capacity. Individual curves represent the rolled shapes commonly used for beams. Figure B.1 is a reproduction of 2 pages (out of 63) from the AISC Manual.

The following example illustrates how designers use such charts to find acceptable shapes for the combined condition.

Example 1. A simple beam of A36 steel carries a total uniformly distributed load of 75 kips on a 36 ft span. Find an acceptable shape for unbraced lengths of (a) 9 ft (b) 12 ft (c) 18 ft.

Solution: First find the maximum bending moment.

$$M = \frac{WL}{8} = \frac{(75)(36)}{8} = 337.5 \text{ kip-ft}$$

Table B.1 yields the following shapes with resisting moments of 337.5 kip-ft or more:

lightest:	W 24 × 76
others:	W 18 × 106
	W 21× 83
	W 27 × 84
	W 30 × 90

Note that Table B.1 also lists the values for L_c and L_u. For the W 24 × 76 L_c is 9.5 ft, so this shape is adequate for part (a).

When the unbraced length falls between L_c and L_u, the allowable bending stress for a W shape drops from 0.66 F_y to 0.60 F_y, a 10 percent loss of of the resisting moment. From Table B.1 you can calculate that the moment capacity of the W 24 × 76 will drop by about 35 kip-ft between unbraced lengths of 9.5 and 11.8 ft, so this shape is not adequate for parts (b) and (c).

In Figure B.1, find the value for the resisting moment on the chart's left edge. The labeled curves all meet this edge at their maximum moment resistance values. Note how the curves are actually horizontal lines until they reach their respective L_c values, where the curves suddenly drop. Eventually they become horizontal lines again—until they reach their respective L_u values.

For this example find a point on the graph that represents the intersection of a moment of 337.5 kip-ft and an unbraced length of 12 ft. Any shape whose curve falls above or to the right of the determined point on the chart is adequate. (The W 24 × 76 curve is left of this point, confirming your earlier calculation of its moment capacity.)

In Figure B.1, the shapes whose curves are solid lines are the lightest members. For this example the curve for a W 24 × 84 is the first solid line above and to the right of the point representing the intersection of 337.5 kip-ft and 12 ft and so is the lightest choice for part (b). Meanwhile the curve for the W 30 × 90 is the first solid line above and to the right of the point representing the intersection of 337.5 kip-ft and 18 ft, and so is the lightest choice for part (c). However, *any* shape to the right is adequate; other possible shapes are

W 14 × 120
W 18 × 97
W 24 × 103
W 27 × 102

It is not always easy to decide whether a beam is sufficiently braced laterally. Figure 6.15 shows a number of common situations in which construction is attached to beams but not all represent sufficient bracing.

In times past, to provide fire protection, designers cast concrete around steel members, as shown in Figure 6.15a. Combining such an element with a cast concrete slab provides a sturdy bracing, although this form of fire protection is no longer common.

Designers frequently use steel beams to support wood joists, as shown in Figure 6.15b. Depending on the form of attachment between the steel and wood elements, such construction may constitute sufficient bracing. Other joist forms—such as the open-web steel (truss) joists shown in Figure 6.15c—are usually attached by welding or bolting, either of which provides adequate bracing. If you bolt a wood member to the top flange of the beam, the bracing is adequate as long as the wood joists are attached to the nailer by metal fasteners (see Figure 6.15d).

When beams support decks directly, you must evaluate how the deck is attached. For example, to ensure adequate bracing, concrete decks are formed to encase the beam flange (see Figure 6.15*e*), have elements welded to the beam and cast into the concrete, or, as with precast decks, have steel elements cast into the deck units and then welded to the steel beams.

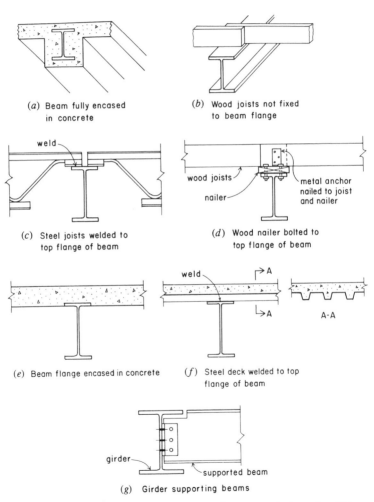

(*a*) Beam fully encased in concrete

(*b*) Wood joists not fixed to beam flange

(*c*) Steel joists welded to top flange of beam

(*d*) Wood nailer bolted to top flange of beam

(*e*) Beam flange encased in concrete

(*f*) Steel deck welded to top flange of beam

(*g*) Girder supporting beams

FIGURE 6.15 Lateral bracing for beams fromsupported construction.

For beams that support other beams, the intersecting supported beams provide significant bracing for the supporting beams. Although the usual type of connection does not grasp either beam flange (see Figure 6.15g), it generally has sufficient twisting resistance to brace against buckling. Therefore, the unbraced length of the supporting beam equals the distance between supported beams. *Note:* Decking also may brace the supporting beams, but since the decking is usually directly supported by and attached to the other beams, designers usually discount such bracing.

In the following problems, which require use of Table B.1 and Figure B.1, choose beams when lateral bracing is a concern.

Problem 6.5.A. A W shape of A36 steel is to be used for a uniformly loaded simple beam carrying a total of 77 kips on a 45 ft span. Select the lightest shape for unbraced lengths of (a) 10 ft (b) 15 ft (c) 22.5 ft.

Problem 6.5.B. A W shape of A36 steel is to be used for a uniformly loaded simple beam carrying a total of 72 kips on a 30 ft span. Select the lightest shape for unbraced lengths of (a) 6 ft (b) 10 ft (c) 15 ft.

6.6 SAFE LOAD TABLES

The simple beam with uniformly distributed load occurs so frequently that it is useful to know a quick way to select shapes based on beam load and span only. The AISC Manual (Ref. 3) provides a series of tables that list data for the W, M, S, and C shapes used most often as beams.

The following example demonstrates how to use such tables. *Note:* The data is the same as for Example 1 of Section 6.2, permitting you to compare design processes.

Example 1. Design a simply supported beam to carry a superimposed load of 2 kips per ft [29.2 kN/m] over a span of 24 ft [7.3 m]. The allowable bending stress is 24 ksi [165 MPa].

Solution: In Section 6.2 I computed the maximum bending moment. With Table B.1, you can use the moment value alone to make a selection. *Note:* Table B.1 also lists L_c and L_u, which you can use to check whether the laterally unbraced length is a problem.

From the table on page 2-61 of the AISC Manual (Ref. 3), meanwhile, identify the row containing a span of 24 ft and then read across for a load of slightly more than 48 kips to allow for the beam weight; this search yields a choice for the W 18 × 46, with a safe load of 52 kips.

So far, the table saves you little time, merely eliminating a single computation—for the bending moment. However, the table contains other data you may use when designing the beam:

For Laterally Unbraced Length. The table gives values for L_c and L_u.

For Deflection: In its rightmost column, the table lists the deflection per span. Note that the beam in question deflects 0.79 in. under the table load. In the example, if the total load is slightly less than 52 kips, the deflection is proportionally less.

For Beam Shear: Listed at the bottom of each column, V is the maximum allowable shear, based on full yield at the limiting shear stress of 40 percent of F_y (see Section 6.3).

For End Bearing: Listed at the bottom of each column, R is the maximum allowable end reaction force with a bearing plate that is 3.5 in. long (from the beam's end). *Note:* The other values grouped at the bottom of each column are for determining the safe reaction for other bearing lengths, as well as for bearing conditions within the beam length (instead of at the beam end). I discuss bearing in Section 6.10.

In sum, these tables are undoubtedly useful design aids for beams. The AISC Manual boasts 104 pages of these tables, including separate sets for F_y values of 36 ksi and 50 ksi.

6.7 EQUIVALENT LOAD TECHNIQUES

The safe service loads listed in the AISC Manual are uniformly distributed loads on simple beams. But because the values are based

ing moment and limiting bending stress, you can refer to the table for other loading conditions. For example, you can use the table when designing framing systems, which always include some beams with other than simple uniformly distributed loadings.

Consider a beam carrying two equal concentrated loads at the third points (Figure B.2, Case 3). For this condition, the figure yields a maximum moment value expressed as $PL/3$. By equating this value to the moment value for a uniformly distributed load, you can derive a relationship between the two loads:

$$\frac{WL}{8} = \frac{PL}{3} \text{ and } W = 2.67 \times P$$

In other words, if you multiply one of the concentrated loads by 2.67, the result is an *equivalent uniform load* (EUL) that produces the same loading as the true loading condition.

Note: In this book I often refer to equivalent uniform load as *equivalent tabular load* (ETL). Figure B.2 yields the ETL factors for several common loading conditions.

Remember that the ETL is based only on flexure (i.e., limiting bending stress); investigations for deflection, shear, or bearing must use the beam's true loading conditions.

You may use this method for any loading condition, not just simple, symmetrical conditions: first find the true maximum moment due to the actual loading and then equate this value to the expression for the maximum moment of a theoretical uniform load. Thus

$$M = \frac{WL}{8} \text{ and } W = \frac{8M}{L}$$

The expression $W = 8M/L$ is the general expression for an equivalent uniform load for any loading condition.

6.8 TORSIONAL EFFECTS

In some situations steel beams experience torsional twisting in addition to shear and bending. For example, torsional twisting exists when the beam is loaded in a plane that does not coincide with the *shear center* of the member cross section. For doubly symmetrical forms, such as the W, M, and S shapes, the shear center coincides with the member centroid (i.e., the intersection of its principal axes). Off-center loadings lead to twisting (see Figure 6.16).

In Figure 6.17, a loading exactly coinciding with the plane of the minor axis (the y-y axis) produces a pure bending about the major axis (the x-x axis). The opposite is also true. However, any misalignment of the loading produces a twisting, torsional effect. For a torsionally weak member—such as open, I shaped sections with thin parts and low bending resistance on the y-y axis—the torsional action often leads to the beam's ultimate failure.

For shapes that are not symmetrical about both axes—such as C and angle shapes—the shear center does not coincide with the member centroid. For example, the channel shape's shear center is some distance behind the back of the section (see Figure 6.17c). Thus, loading the channel through its centroid or its vertical web portion can produce twisting. One way to avoid this torsional action is to attach an angle to the channel, permitting the load to be applied close to the channel's shear center.

Figure 6.18 shows the locations of shear centers and centroids for various shapes and combinations of shapes. When these two centers coincide (see Figure 6.18a), designers' concern is limited to assuring that loadings remain in the plain of the member's centroidal axis. When the two centers are separated (see Figure 6.18b), centroidal loading produces twisting unless the plane of loading

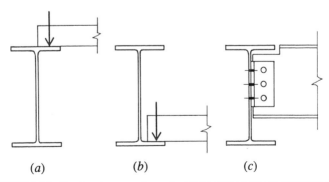

     (*a*)   (*b*)   (*c*)

FIGURE 6.16 Torsional moment on a beam produced by off-center loading.

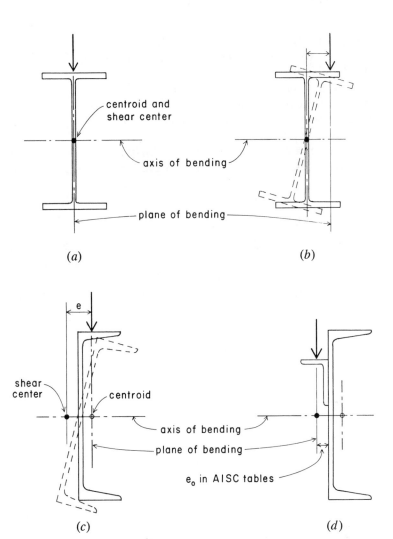

(a)   *(b)*

(c)   *(d)*

FIGURE 6.17 Torsion on beams developed by loading not aligned with the beam's shear center.

also passes through the shear center (see Figure 6.18c, which shows the single tee shape and channel, and Figure 6.18d, which shows the double-angle and double-channel).

Figure 6.19a shows the torsional buckling that may occur in beams with low torsional resistance. If torsional buckling occurs in tandem with some other torsional effect, a beam may lose all its resistance to load.

Because designers can calculate torsional effects, they can design members to resist torsion. However, for beams in structural

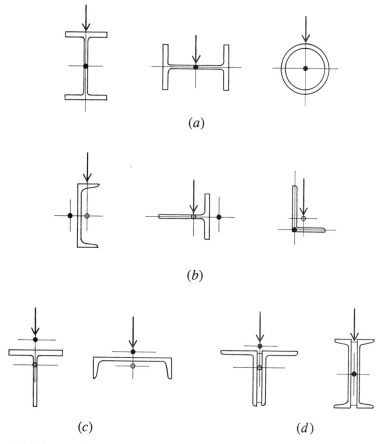

(a)

(b)

(c)   (d)

FIGURE 6.18 Shear centers and centroids for various beam sections. For all cases the loading passes through the centroid.

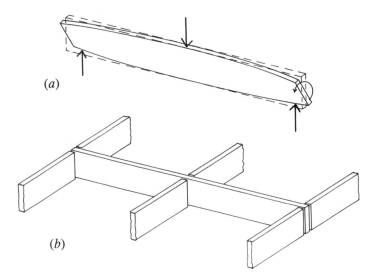

(*a*)

(*b*)

FIGURE 6.19 The bracing effect of intersecting framing (prevents torsional failure of a beam).

frameworks, designers prefer to avoid torsional effects as much as possible. To do so, they may provide bracing—for example, a common solution with wood joist construction is to place members perpendicular to the affected beam to prevent torsional rotation at the beam's ends and midspan (see Figure 6.19*b*).

In sum, torsion is a basic concern. To ensure safety, you must develop appropriate construction details.

6.9 CONCENTRATED LOAD EFFECTS ON BEAMS

Either an excessive bearing reaction on a beam or an excessive concentrated load at some point in the beam span may cause either localized yielding or *web crippling* (i.e., buckling of the thin beam web). The AISC Specification requires designers to investigate beam webs for such effects; moreover, the AISC Specification requires designers to use web stiffeners if the concentrated load exceeds limiting values.

Figure 6.20 shows three common situations. Figure 6.20*a* shows the beam end bearing on a support (commonly a masonry or con-

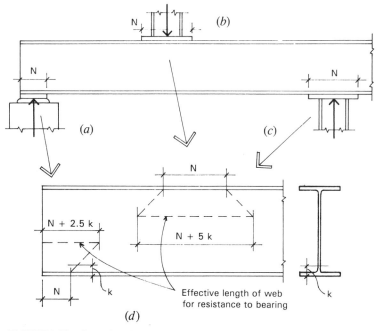

FIGURE 6.20   Considerations for bearing in beams with thin webs, as related to web crippling.

crete wall), with the reaction force transferred to the beam bottom flange through a steel bearing plate. Figure 6.20*b* shows a column load applied to the top of the beam at some point within the beam span. Figure 6.20*c* shows a beam supported in bearing on top of a column with the beam continuous through the joint.

Figure 6.20*d* shows the portion of the web length (along the beam span) that is assumed to resist bearing forces. For yield resistance, the maximum end reaction and the maximum load within the beam span are defined as follows:

Maximum end reaction $= (0.66 \ F_y) \ (t_w) \ \{N \ + \ 2.5(k)\}$
Maximum interior load $= (0.66 \ F_y) \ (t_w) \ \{N \ + \ 5(k)\}$

where t_w = thickness of the beam web
N = length of the bearing

k = distance from the outer face of the beam flange to the web toe of the fillet (radius) of the corner between the web and the flange

Note: Tables in the AISC Manual list dimensions t_w and k for W shapes.

When the maximum end reaction and interior load are exceeded, a beam needs web stiffeners—for example, steel plates welded into the beam's channel-shaped sides (see Figure 6.21)—at the location of the concentrated load. Not only do stiffeners improve bearing resistance, but also they alleviate the potential for web crippling.

Although the AISC Manual provides formulas for computing limiting loads due to web crippling, it also includes data tables that engender shortcut methods.

In the following example I illustrate how to compute a limiting end reaction using formulas, and I demonstrate how to find a solution using data tables from the AISC.

Example 1. A W 18 × 55 beam of A36 steel has an end reaction that is developed in bearing over a length N = 10 in. [254 mm]. Inves-

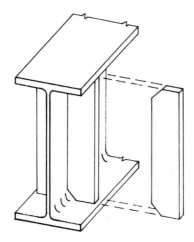

FIGURE 6.21 Use of stiffeners to prevent lateral buckling of a beam's thin web.

tigate the beam for yielding and web crippling if the reaction force is 44 kips [196 kN].

Solution: Table A.1 yields values of 1.3125 in. [33 mm] for k and 0.390 in. [10 mm] for t_w. Calculate the end bearing limit:

$$\text{Maximum } R = (0.66 \ F_y) (t_w) \{N + 2.5(k)\}$$

$$= (24) (0.390) \{10 + 2.5(1.3125)\} = 124 \text{ kips } [538 \text{ kN}]$$

To investigate for buckling (web crippling), I use the following data from the AISC Manual, page 2:61:

$R_1 = 30.4$ kips
$R_2 = 9.27$ kips/in.
$R_3 = 39.4$ kips
$R_4 = 3.18$ kips/in.

The maximum end reaction is the least of two values, one based on yield and the other on buckling:

$$\text{Maximum } R = R_1 + N(R_2) \quad \text{or} \quad R_3 + N(R_4)$$

$$= 30.4 + 10(9.27) = 123.1 \text{ kips (for yield)}$$

$$\text{or, } = 32.6 + 10(3.18) = 71.2 \text{ kips (for buckling)}$$

The buckling limit prevails, so the maximum end reaction is 71.2 kips. Because this value is less than the required force of 44 kips, the beam is adequate as is.

In Figures 6.20*b* and *c*, the beam is continuous through the bearing condition. In such cases you must investigate web crippling precisely, using AISC formulas. However, as with yielding, the crippling limit is slightly higher than that for end bearing. As a result, you may compare the web crippling for end bearing (found from AISC table data) to the required load: if the required load is less than (or only very slightly over) the web crippling, the situation is not critical.

Example 2. The beam in Example 1 carries a column load of 70 kips [311 kN] within the beam span. The bearing length of the column on the beam is 12 in. [305 mm]. Investigate the beam for yield and web crippling.

Solution. First investigate for yield.

$$\text{Maximum } P = (0.66\ F_y)\,(t_w)\,\{N + 5(k)\}$$
$$= (24)\,(0.390)\{12 + 5(1.3125)\} = 173.7 \text{ kips } [773 \text{ kN}]$$

For web crippling consider the limit for end bearing as follows:

$$\text{Maximum } R = R_3 + N(R_4) = 39.4 + 12(3.18) = 77.56 \text{ kips}$$

Because both values exceed the required force, and because the actual crippling resistance within the beam span is higher than the load, the beam is sufficient.

Problem 6.9.A. Find the maximum allowable reaction force for a W 18 × 40 beam with an end bearing plate 8 in. [203 mm] long.

Problem 6.9.B. A column load of 81 kips [360 kN] with a bearing length of 11 inches [279 mm] is placed atop the beam defined in Problem 6.9.A. Are web stiffeners required?

6.10 BEARING PLATES

Beams that are supported on masonry or concrete usually rest on top of steel bearing plates. Although the plates often help connect the beam end to the support, their primary purpose is to provide an extended area of contact bearing in order to lower the bearing pressure to a value sustainable by the bearing support material (see Figure 6.22).

For wall bearing, the dimension N usually depends on the wall thickness—in fact, N is typically about 2 in. less than the wall thickness. If the wall is thin and the bearing load large, you may need a large B. Of course, as B increases, the bending of the plate requires an increase in the plate thickness, t. At some point, B becomes too large for the general form of the connection, and you must modify the structure (thicker wall, more beams to decrease the end reaction, and so on).

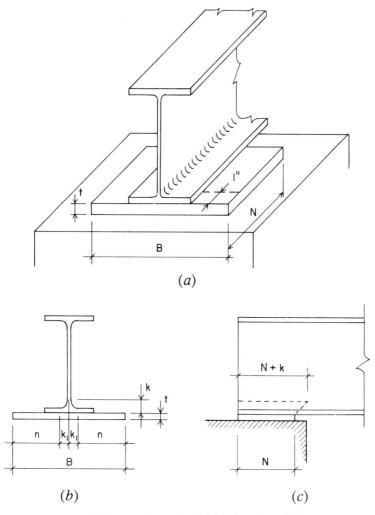

FIGURE 6.22 Reference dimensions for bearing plates for beam ends.

To determine the required thickness for the bearing plate, consider the bending generated on the cantilever distance labeled n in Figure 6.22b. The load on this cantilever is the uniformly distributed pressure on the bottom of the plate. Based on this bending moment, you can derive a formula for the required plate thickness:

$$t = \sqrt{\frac{3f_p n^2}{F_b}}$$

where t = thickness of the plate

f_p = computed bearing pressure

F_b = allowable bending stress for the plate—usually 0.75 F_y, or 27 ksi for A36 steel

n = $(B/2) - k_1$

k_1 = distance from the center of the beam web to the toe of the flange fillet

The AISC table of properties lists k_1 for W shapes. Table 6.1 lists common values used for bearing on masonry and concrete walls.

You also must investigate the beam for yield and web crippling, as I discussed in Section 6.9. If this investigation indicates that web stiffeners are required, the bending of the bearing plate is slightly reduced. With stiffeners, you can reduce the cantilever to the distance beyond the flange's edge.

In the following example I illustrate how to design a bearing plate like the one shown in Figure 6.22a.

TABLE 6.1 Allowable Bearing Pressure on Concrete and Masonry

Type of Material and Conditions	Allowable Unit Stress in Bearing F_p	
	psi	kPa
Solid brick, reinforced, type S mortar		
f'_m = 1500 psi	170	1200
f'_m = 4500 psi	338	2300
Hollow unit masonry (CMU), type S mortar		
f'_m = 1500 psi	225	1500
(on net area of masonry)		
Concrete[a]		
(1) Bearing on full area of support		
f'_c = 3000 psi	750	5000
f'_c = 4000 psi	1000	7000
(2) Bearing on 1/3 of support area or less		
f'_c = 3000 psi	1125	7500
f'_c = 4000 psi	1500	10000

[a]Allowable stresses for areas between these limits may be determined by direct proportion.

Example 1. A beam consisting of a W 21 × 57 transfers an end reaction force of 44 kips [196 kN] to a solid brick wall via an A36 steel bearing plate. Assume a value of $f'_m = 1500$ for the masonry and a limiting dimension perpendicular to the wall (N in Figure 6.22) of 10 in [254 mm]. Select the plate thickness and its other dimension (B in Figure 6.21).

Solution: From Table A.1, $k_1 = 0.875$ in. [22 mm]. From Table 6.1, the allowable bearing pressure, F_p, is 170 psi [1200 kPa]. For the required area of the plate

$$A = \frac{R}{F_p} = \frac{44{,}000}{170} = 259 \text{ in.}^2 \ [163 \times 10^3 \text{ mm}^2]$$

Because $N = 10$ in. [254 mm],

$$B = \frac{259}{10} = 25.9 \text{ in. [643 mm]}$$

rounded off to 26 in.

Now determine the true pressure:

$$f_p = \frac{44{,}000}{10 \times 26} = 169 \text{ psi [1187 kPa]}$$

To find the plate thickness, first determine n:

$$n = \frac{B}{2} - k_1 = \frac{26}{2} - 0.875 = 12.125 \text{ in. [303 mm]}$$

In turn, the required thickness is

$$t = \sqrt{\frac{3f_p n^2}{F_b}} = \sqrt{\frac{3 \times 169 \times (12.125)^2}{27{,}000}}$$

$$= \sqrt{2.2760} = 1.66 \text{ in. [42 mm]}$$

rounded up to 1.75 in.

In sum, the plate is specified as 10 in. × 2 ft − 2 in. × 1.75 in.

In some situations, depending on the beam flange width, a low bearing pressure under the reaction force permits designers to install the beam without a bearing plate. Where this is true, be sure to investigate the bending in the beam flange: if bending stress is excessive, use a plate to stiffen the beam flange.

Problem 6.10.A. A W 14 × 30 beam with a reaction of 20 kips [89 kN] rests on a brick wall with brick of f'_m = 1500 psi and Type S mortar. The beam has an A36 steel bearing plate with a length of 8 in. [203 mm] parallel to the beam length. Determine the plate's other dimensions.

6.11 MANUFACTURED TRUSSES

Parallel-chord trusses are produced in many sizes by many manufacturers. Most producers comply with industry-wide regulations. For light steel trusses the principal such organization is the Steel Joist Institute (SJI). Although SJI publications are a chief source of general information (see Ref. 7), the products of individual manufacturers vary enough that you should consult the supplier of a specific product for design data.

Early versions of light steel parallel-chord trusses, called *open-web joists*, used steel bars for the chords and the continuously bent web members (see Figure 6.23); in fact, those trusses were known as *bar joists*. Although other elements now are used for the chords, the bent steel rod is still a popular element for some of the smaller joists.

Some members are as long as 150 ft and have depths of more than seven feet. Larger members usually take the forms common

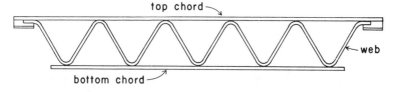

FIGURE 6.23 Form of a short-span open-web steel joist.

TABLE 6.2 Safe Service Loads for K-Series Open Web Joists[a]

Joist Designation	12K1	12K3	12K5	14K1	14K3	14K6	16K2	16K4	16K6	18K3	18K5	18K7	20K3	20K5	20K7
Weight (lb/ft)	5.0	5.7	7.1	5.2	6.0	7.7	5.5	7.0	8.1	6.6	7.7	9.0	6.7	8.2	9.3
Span (ft)															
20	241 (142)	302 (177)	409 (230)	284 (197)	356 (246)	525 (347)	368 (297)	493 (386)	550 (426)	463 (423)	550 (490)	550 (490)	517 (517)	550 (550)	550 (550)
22	199 (106)	249 (132)	337 (172)	234 (147)	293 (184)	432 (259)	303 (222)	406 (289)	498 (351)	382 (316)	518 (414)	550 (438)	426 (393)	550 (490)	550 (490)
24	166 (81)	208 (101)	282 (132)	196 (113)	245 (141)	362 (199)	254 (170)	340 (221)	418 (269)	320 (242)	434 (318)	526 (382)	357 (302)	485 (396)	550 (448)
26				166 (88)	209 (110)	308 (156)	216 (133)	289 (173)	355 (211)	272 (190)	369 (249)	448 (299)	304 (236)	412 (310)	500 (373)
28				143 (70)	180 (88)	265 (124)	186 (106)	249 (138)	306 (168)	234 (151)	318 (199)	385 (239)	261 (189)	355 (248)	430 (298)
30							161 (86)	216 (112)	266 (137)	203 (123)	276 (161)	335 (194)	227 (153)	308 (201)	374 (242)
32							142 (71)	190 (92)	233 (112)	178 (101)	242 (132)	294 (159)	199 (126)	271 (165)	328 (199)
36										141 (70)	191 (92)	232 (111)	157 (88)	213 (115)	259 (139)
40													127 (64)	172 (84)	209 (101)

Joist Designation	22K4	22K6	22K9	24K4	24K6	24K9	26K5	26K6	26K9	28K6	28K8	28K10	30K7	30K9	30K12
Weight (lb/ft)	8.0	8.8	11.3	8.4	9.7	12.0	9.8	10.6	12.2	11.4	12.7	14.3	12.3	13.4	17.6
Span (ft)															
28	348 (270)	427 (328)	550 (413)	381 (323)	467 (393)	550 (456)	466 (427)	508 (464)	550 (501)	548 (541)	550 (543)	550 (543)			
30	302 (219)	371 (266)	497 (349)	331 (262)	406 (319)	544 (419)	405 (346)	441 (377)	550 (459)	477 (439)	550 (500)	550 (500)	550 (543)	550 (543)	550 (543)
32	265 (180)	326 (219)	436 (287)	290 (215)	357 (262)	478 (344)	356 (285)	387 (309)	519 (407)	418 (361)	515 (438)	549 (463)	501 (461)	549 (500)	549 (500)
36	209 (126)	257 (153)	344 (201)	229 (150)	281 (183)	377 (241)	280 (199)	305 (216)	409 (284)	330 (252)	406 (306)	487 (366)	395 (323)	475 (383)	487 (392)
40	169 (91)	207 (111)	278 (146)	185 (109)	227 (133)	304 (175)	227 (145)	247 (157)	331 (207)	266 (183)	328 (222)	424 (284)	319 (234)	384 (278)	438 (315)
44	139 (68)	171 (83)	229 (109)	153 (82)	187 (100)	251 (131)	187 (100)	204 (118)	273 (155)	220 (137)	271 (167)	350 (212)	263 (176)	317 (208)	398 (258)
48				128 (63)	157 (77)	211 (101)	157 (83)	171 (90)	229 (119)	184 (105)	227 (128)	294 (163)	221 (135)	266 (160)	365 (216)
52							133 (66)	145 (80)	195 (102)	157 (83)	193 (100)	250 (128)	188 (106)	226 (126)	336 (184)
56										135 (66)	166 (80)	215 (102)	162 (84)	195 (100)	301 (153)
60													141 (69)	169 (81)	262 (124)

[a] Loads in pounds per ft of joist span. First entry represents the total joist capacity; entry in parentheses is the load that produces a deflection of 1/360 of the span. See Figure 6.24 for definition of span.

Source: Data adapted from more extensive tables in the *Standard Specifications, Load Tables, and Weight Tables for Steel Joists and Joist Girders,* 1994 (Ref. 7), with permission of the publishers, Steel Joist Institute. The Steel Joist Institute publishes both specifications and load tables; each of these contains standards that are to be used in conjunction with one another.

for steel trusses—double-angles, structural tees, and so on. (Refer to the SJI, as well as individual suppliers, for more information regarding installation details, suggested specifications, bracing, and safety during erection.) Smaller members are used for both floor joists and roof rafters.

Table 6.2, adapted from a standard SJI table, lists several sizes available in the K series, which is the lightest group of joists. Note that joists are identified by a three-unit designation: the first number indicates the joist's overall nominal depth, the letter indicates the series, and the second number indicates the size of the joist's members (the higher the number, the heavier and stronger the joist).

You can use Table 6.2 to select the proper joist for a determined load and span situation. Each span usually has two entries in the table; the first number represents the joist's total load capacity (lb/ft) and the number in parentheses is the load that produces a deflection of 1/360 of the span.

Example 1. Use open-web steel joists to support a roof with a unit live load of 20 psf and a unit dead load of 15 psf (not including the weight of the joists) on a span of 40 ft. Joists are spaced at 6 ft center to center, Select the lightest joist if deflection under live load is limited to 1/360 of the span.

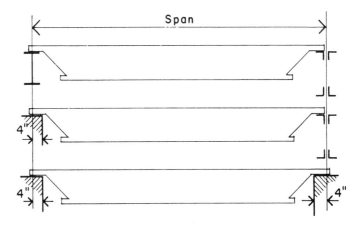

FIGURE 6.24 Definition of span for open-web steel joists, as given in Ref. 7. Reprinted by permission of the Steel Joist Institute.

Solution: First determine the unit load per ft on the joists:

Live load = 6(20) = 120 lb/ft
Dead load = 6(15) = 90 lb/ft
Total load = 120 + 90 = 210 lb/ft

With total load and live load values in hand, scan the data in Table 6.2 for the given span. Note that the joist weight—so far excluded from the computation—is included in the table's total load entries. Once you select a joist, deduct the actual joist weight from the table entry before comparing with the computed values.

Table 6.3 lists the possible choices; the 24K6 is the lightest choice.

Example 2. Use open-web steel joists for a floor with a unit live load of 75 psf and a unit dead load of 40 psf (not including the joist weight) on a span of 30 ft. Joists are 2 ft on center. Deflection is limited to 1/240 of the span under total load and 1/360 of the span under live load only. Find the lightest joist possible and the lightest joist of least depth possible.

Solution: As in the previous example, first determine the unit loads:

Live load = 2(75) = 150 lb/ft
Dead load = 2(40) = 80 lb/ft
Total load = 150 + 80 = 230 lb/ft

TABLE 6.3   Possible Choices for the Roof Joist

| Load Condition | Load per Foot for the Indicated Joists | | | |
	22K9	24K6	26K6	28K6
Total capacity, from Table 6.2	278	227	247	266
Joist weight, from Table 6.2	11.3	9.7	10.6	11.4
Net usable capacity	266.7	217.3	236.4	254.6
Load for deflection of 1/360, from Table 6.2	146	133	157	183

To satisfy the deflection criteria for total load, the limiting value for deflection should be not less than $(240/360)(230) = 153$ lb/ft. Since this value is slightly larger than the live load, it is your reference point.

Table 6.4 lists the possible choices; the lightest joist is the 22K4 and the shallowest depth joist is the 18K5.

In some situations you may want to select a deeper joist, even though its load capacity is somewhat redundant. For example, total sag, rather than an abstract curvature limit, may be more significant for a flat roof structure. In Example 1, a 1/360 sag of the 40 ft span $= (1/360)(40 \times 12) = 1.33$ in. You must consider how such a sag limits roof drainage or affects interior partition walls. Another example: Designers sometimes deliberately choose the deepest feasible joist for floor structures to reduce deflection and thereby stiffen the structure in general against bouncing effects.

Stability is a major concern because joists have very little lateral or torsional resistance. Other construction elements, such as decks and ceiling framing, may help, but you must study the whole bracing situation carefully. In general, lateral bracing in the form of x-braces or horizontal ties is required for all steel joist construction.

One way to improve stability involves the typical end support detail, shown in Figure 6.23. Hanging trusses by the ends of their top chords, which avoids the rollover type of rotational buckling at the supports illustrated in Figure 6.14*b*, adds a dimension to the overall construction depth. This added dimension (the depth of the end of the joist) is typically 2.5 in. for small joists and 4 in. for larger joists.

TABLE 6.4   Possible Choices for the Floor Joist

Load Condition	Load per Foot for the Indicated Joists		
	18K5	20K5	22K4
Total capacity, from Table 6.2	276	308	302
Joist weight, from Table 6.2	7.7	8.2	8.0
Net usable capacity	268.3	299.8	294
Load for deflection of 1/360, from Table 6.2	161	201	219

When developing a complete truss system, most designers use a special type of prefabricated truss: the *joist girder*. This truss is designed specifically to carry the regularly spaced, concentrated loads consisting of the end support reactions of joists. Figure 6.25 shows a common joist girder form. Joist girder designations follow a standard form: nominal girder depth, number of spaces between joists (called the girder *panel unit*), and end reaction force from the joists (i.e., the unit concentrated load on the girder).

You can select a joist girder from catalogs. The procedure is usually as follows:

1. Determine the joist spacing, joist load, and girder span. (The joist spacing should be a full number division of the girder span.)

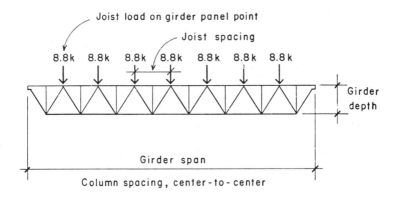

Standard Designation:

48 G	8 N	8.8 K
Depth in inches	Number of joist spaces	Load on each panel point in kips

Specify: 48G8N8.8K

FIGURE 6.25 Considerations for layout and designation of joist girders.

2. Identify the girder, using the standard designation.
3. Choose the girder from a manufacturer's catalog, or specify what you need to a supplier.

I further discuss the use of joists and complete truss systems in the building design examples in Chapter 12.

Problem 6.11.A. Use open-web steel joists for a roof with a live load of 25 psf and a dead load of 20 psf (not including the joist weight) on a span of 48 ft. Joists are 4 ft on center. Deflection under live load is limited to 1/360 of the span. Select the lightest joist.

Problem 6.11.B. Use open-web steel joists for a roof with a live load of 30 psf and a dead load of 18 psf (not including the joist weight) on a span of 44 ft. Joists are 5 ft on center. Deflection is limited to 1/360 of the span. Select the lightest joist.

Problem 6.11.C. Use open-web steel joists for a floor with a live load of 50 psf and a dead load of 45 psf (not including the joist weight) on a span of 36 ft. Joists are 2 ft on center. Deflection is limited to 1/360 of the span under live load only and 1/240 of the span under total load. Select (a) the lightest possible joist (b) the shallowest joist possible.

Problem 6.11.D. Repeat Problem 6.11.C, except use the following data: live load is 100 psf, the dead load is 35 psf, and the span is 26 ft.

6.12   STEEL DECKS

Steel decks consisting of formed sheet steel are produced in a variety of configurations, as shown in Figure 6.26. The simplest is the corrugated sheet, shown in Figure 6.26a, which may be used as the single surface for walls and roofs of utilitarian buildings (tin shacks) or as the surfacing of a built-up panel or general sandwich-type construction. As a structural deck, the simple corrugated sheet is used for very short spans, typically with a structural-grade concrete fill that serves as the spanning deck.

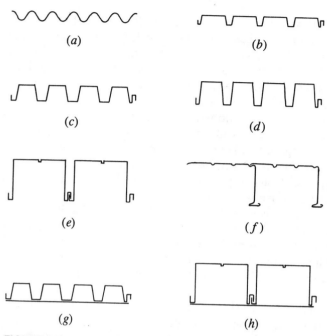

FIGURE 6.26   Cross section shapes of formed sheet steel deck units.

Figure 6.26*b* to *d* shows three variations of a widely used product. When used for roofs, where loads are light, a flat top surface is formed with a lightweight fill of foamed concrete, a light fill of gypsum concrete, or a rigid sheet material. For floors, where loads are heavier, a structural-grade concrete fill is used; designers often select deeper ribs (see Figure 6.26*c* and *d*) to achieve greater spans with widely spaced beams. Common overall deck heights are 1.5, 3, and 4.5 in.

Formed sheet steel decks produced with considerable depth, such as those shown in Figure 6.26*e* and *f*, can achieve considerable span, generally combining the functions of joists and deck.

Sometimes steel deck units serve as conduits for power, signal, or communication wiring. In such cases designers close the deck cells with a flat sheet of steel (see Figures 6.26*g* and *h*) to provide for wiring in one direction in a grid; the perpendicular wiring is provided in conduits buried in the concrete fill.

Decks vary in form (cross section shape) and thickness (gage). Designers usually base their choice of decks on load and span conditions. Units are typically available in lengths of 30 ft or more; such length permits design for a multiple-span condition. Although using decks reduce bending effects only slightly, such a design reduces deflection and bouncing a great deal.

Fire protection for floor decks is provided partly by the concrete fill. To protect the underside, designers use either sprayed on fireproofing materials or a permanent fire-rated ceiling construction. The latter solution is no longer favored, however, because many disastrous fires have started in the void space between ceilings and overhead floors or roofs.

Designers also use the deck as a horizontal diaphragm for distributing lateral forces from wind and earthquakes, and as lateral bracing for beams and columns.

When structural-grade concrete is used as a fill, its relationship to a forming steel deck takes one of three forms:

- The concrete serves strictly as a structurally-inert fill, providing a flat surface, fire protection, added acoustic separation, and so on. It makes no significant structural contribution.
- The steel deck functions essentially only as a forming system for the concrete fill—in other words, the concrete is reinforced and designed as a spanning structural deck.
- The concrete and sheet steel work together in *composite structural action*. In effect, the sheet steel serves as reinforcement for midspan bending stresses. (Of course, the fill still needs top reinforcement for negative bending moments over the deck supports.)

Table 6.5 presents data adequate for preliminary design work; the data relates to the use of the deck unit in Figure 6.26*b* for roof structures. *Note:* For roof decks with such units, structural integrity depends solely on the steel deck, not the concrete fill.

Three different rib configurations—described as *narrow, intermediate*, and *wide*—are shown for the deck units in Table 6.5. Because each configuration has some effect on the properties of the deck cross section, the table includes three sections. Note that structural peformance is a minor factor when choosing rib width.

TABLE 6.5 Safe Service Load Capacity of Formed Steel Roof Deck

Deck [b] Type	Span Condition	Weight [c] (psf)	Total (Dead & Live) Safe Load [d] for Spans Indicated in ft-in.												
			4-0	4-6	5-0	5-6	6-0	6-6	7-0	7-6	8-0	8-6	9-0	9-6	10-0
NR22	Simple	1.6	73	58	47										
NR20		2.0	91	72	58	48	40								
NR18		2.7	121	95	77	64	54	46							
NR22	Two	1.6	80	63	51	42									
NR20		2.0	96	76	61	51	43								
NR18		2.7	124	98	79	66	55	47	41						
NR22	Three or More	1.6	100	79	64	53	44								
NR20		2.0	120	95	77	63	53	45							
NR18		2.7	155	123	99	82	69	59	51	44					
IR22	Simple	1.6	86	68	55	45									
IR20		2.0	106	84	68	56	47	40							
IR18		2.7	142	112	91	75	63	54	46	40					
IR22	Two	1.6	93	74	60	49	41								
IR20		2.0	112	88	71	59	50	42							
IR18		2.7	145	115	93	77	64	55	47	41					
IR22	Three or More	1.6	117	92	75	62	52	44							
IR20		2.0	140	110	89	74	62	53	46	40					
IR18		2.7	181	143	116	96	81	69	59	52	45	40			
WR22	Simple	1.6			(89)	(70)	(56)	(46)							
WR20		2.0			(112)	(87)	(69)	(57)	(47)	(40)					
WR18		2.7			(154)	(119)	(94)	(76)	(63)	(53)	(45)				
WR22	Two	1.6			98	81	68	58	50	43					
WR20		2.0			125	103	87	74	64	55	49	43			
WR18		2.7			165	137	115	98	84	73	65	57	51	46	41
WR22	Three or More	1.6			122	101	85	72	62	54	(46)	(40)			
WR20		2.0			156	129	108	92	80	(67)	(57)	(49)	(43)		
WR18		2.7			207	171	144	122	105	(91)	(76)	(65)	(57)	(50)	(44)

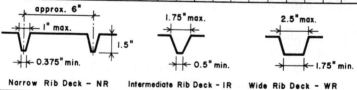

approx. 6"
←1" max.
←0.375" min.
1.5"

1.75" max.
←0.5" min.

2.5" max.
←1.75" min.

Narrow Rib Deck – NR   Intermediate Rib Deck – IR   Wide Rib Deck – WR

[a] Letters refer to rib type (see illustrations). Numbers indicate gage (thickness) of deck sheet steel.

[b] Approximate weight with paint finish; other finishes available.

[c] Total safe allowable service load in lb/sq ft. Loads in parentheses are governed by live load deflection not in excess of 1/240 of the span, assuming a dead load of 10 lb/sq ft.

Source: Adapted from the *Steel Deck Institute Design Manual for Composite Decks, Form Decks, and Roof Decks*, (Ref. 8), with permission of the publishers, the Steel Deck Institute.

For example, more important is whether the deck is to be welded to its supports (usually required for good diaphragm action); if so, a wide rib is required. Another example: If a relatively thin topping material is used, the narrow rib is favored. Rusting is a critical problem for the very thin sheet steel deck. Although the deck's top is usually protected by other construction, its underside requires treatment. Because many manufacturers add an appropriate surfacing in the factory, the deck weights in Table 6.5 take into account paint, which is usually the least expensive rust-proofing treatment. Bonded enamel or galvanizing surfaces add to the deck weight.

Although these products are typically available in lengths of 30 ft or more, continuity depends on the spacing of supports. Table 6.5 provides for three continuity cases: simple span (one span), two span, and three or more spans.

Problem 6.12.A,B,C. Using data from Table 6.5, select the lightest steel deck.

A. Simple span of 7 ft, total load of 45 psf.
B. Two-span condition, span of 8.5 ft, total load of 45 psf.
C. Three-span condition, span of 6 ft, total load of 50 psf.

6.13 ALTERNATIVE DECKS WITH STEEL FRAMING

Figure 6.27 shows four possible floor decks for a framing system of rolled steel beams.

The wood deck (Figure 6.27a) is usually supported by and nailed to a series of wood joists, which are in turn supported by the steel beams. However, in some cases the deck may be nailed to wood members that are bolted to the tops of the steel beams, as shown in the figure. For floor construction, designers often use a concrete fill on top of the wood deck to add stiffness, increase fire protection, and improve acoustic behavior.

With a sitecast concrete deck (Figure 6.27b), designers typically place plywood panels against the bottoms of the beams' top flanges to lock the slab and beams together for lateral effects; in addition, steel lugs are welded to the tops of the beams for composite con-

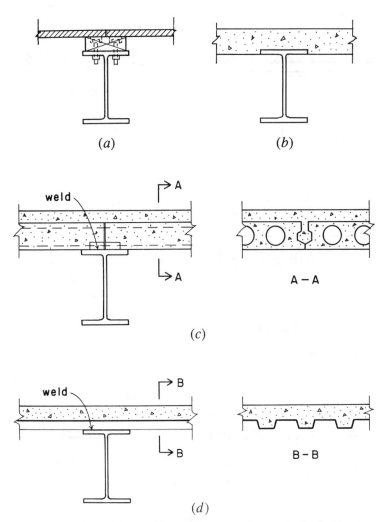

FIGURE 6.27 Typical forms of floor deck construction used with steel framing.

struction (see Section 9.2). For precast deck units, steel elements are embedded in the ends of the precast units and welded to the beams. Designers typically use a site-poured concrete fill to provide a smooth top surface; this fill is bonded to the precast units to improve structural performance.

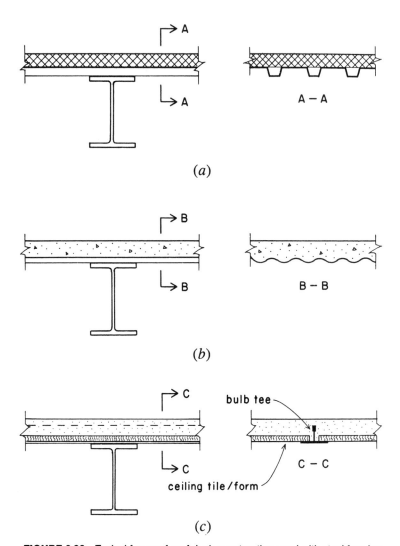

FIGURE 6.28 Typical forms of roof deck construction used with steel framing.

As I described in Section 6.12, formed sheet steel units may be used in one of three ways: as the primary structure, as forming for the concrete deck, or as a composite element in conjunction with the concrete. To attach this type of deck to the steel beams, designers usually weld the steel units to the beams before the concrete is placed.

Figure 6.28 shows roof decks that use steel elements. Roof loads are typically lighter than floor loads, and bounce is not a major concern unless you want to hang elements from the deck; such a system is vulnerable to vertical movements during an earthquake. A fourth possibility is the plywood deck in Figure 6.27*a*.

7

STEEL COLUMNS

Steel columns range from single-piece members (pipe, tube, W shape, etc.) to built-up assemblages; the ordinary column form is a vertical, linear member of constant cross section shape. The column exists essentially to resist an axial compressive force; however, many columns also must sustain bending, shear, torsion, and even tension under varying loading conditions.

In this chapter I discuss the general use of steel columns, with an emphasis on the common, simple, single-piece column.

7.1 INTRODUCTION

A column or strut is a compression member whose length is several times its width (or thickness). The term *column* is reserved for relatively tall, vertical supports, but the term *strut* sometimes refers to other forms of compression members, such as braces or truss members. *Note:* Other compression members are pilasters, piers, and pedestals, although these occur mostly in concrete or masonry construction.

For the linear column, with a compression force applied axially (coinciding at all points with the centroid of the column cross section), the average unit stress on the cross section is expressed as

$$f = \frac{P}{A}$$

where f = average unit compressive stress, uniformly distributed on the cross section

P = axial compressive force

A = area of the column cross section

Consider a small block of steel 1×1 in. in cross section and 2 in. high (see Figure 7.1a). When subjected to compression, the block develops a compressive stress all the way into the material's yield range; thus the limit for the force P is based on the material's stress limit. If, however, a member with the same cross section is considerably longer—say, 30 to 40 in., as shown in Figure 7.1b—the maximum compressive resistance may be less than that of the block because the more slender element tends to fail by buckling sideways before the material's full stress resistance is reached.

The graph in Figure 7.2 shows the complete range of column behavior, with considerations for slenderness as well as stress. *Note:* This graph is actually a plot of the range of allowable stresses for columns with F_y = 36 ksi and 50 ksi.

From the graph you can readily identify three basic forms of behavior. In the first zone, the predominant response is stress-related, and ultimate failure is essentially limited by the yield of the material, with allowable stress limited to some percentage of the yield strength ($0.60\,F_y$ according to the AISC specification). In the third zone, where the very slender column fails by buckling, the load limit is established by the classical Euler buckling load formula for *elastic buckling* conditions. Between these zones is an intermediate response zone, called the *inelastic buckling range*, where the member begins to stiffen and resist buckling.

When designing a column, you must consider many factors, including material and cross section shape. You also must take into account the general usage conditions, based on the development of the total building framework. For the final structural design, you must determine the design loads, along with any conditions relat-

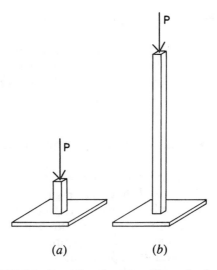

FIGURE 7.1 Examples of varying column slenderness.

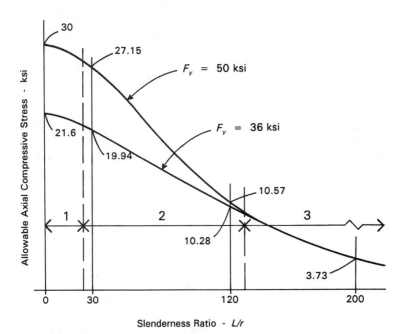

FIGURE 7.2 Allowable axial compressive stress for columns as a function of yield limit and column slenderness. Range 1 covers a yield stress failure condition. Range 3 covers an elastic buckling limit based on steel stiffness and independent of steel yield strength. Range 2 covers the *inelastic buckling* condition.

ing to end restraints or lateral bracing. In addition, keep in mind whether other structural actions—bending, shear, torsion—will combine with the compressive action. The final column form and dimensions must reflect all pertinent factors.

7.2 COLUMN FORMS WITH STEEL PRODUCTS

The most common building column forms are the round pipe, the rectangular tube, and the W shape (see Figure 7.3). Pipe and tube columns usually are used only for one-story buildings, whereas W shape columns frequently are extended through several stories.

Supported framing typically sits atop pipe and tube columns; it occasionally does so with W shape columns. The W shape column is especially useful when framing must be attached to the *sides* of columns, as with multistory construction.

The most common W shapes have nominal depths of 10, 12, or 14 in. In addition, the popular W shapes have relatively wide flanges. Designers prefer columns with such depth and width to handle framing in two directions; in other words, designers require columns with a sufficient depth to accommodate beams from one direction and sufficient flange width to handle beams from the other direction.

Pipes and tubes are ideal for resisting axial compression, but the W shape has a weak axis (y-y) and a strong axis (x-x). However, the W shape is especially useful if you can orient its strong bending axis in response to a major bending force that is combined with the column compression load.

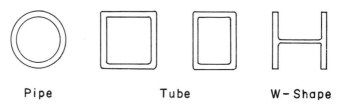

Pipe   Tube   W - Shape

FIGURE 7.3 Common steel columns cross section shapes.

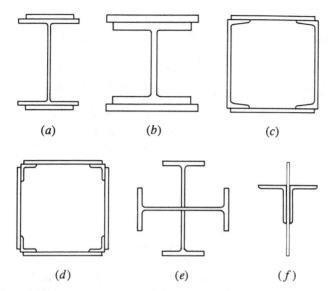

(a) (b) (c)

(d) (e) (f)

FIGURE 7.4 Forms of combined, built-up column shapes.

For various reasons, sometimes it is necessary to make a column section by assembling two or more elements. Figure 7.4 shows some common assemblages that are used when a particular size or shape is not available or where some special structural task is required. In the past, designers needed built-up sections because it was difficult to find large rolled shapes. In recent years, as manufacturers made more large elements, designers have not had to spend money fabricating built-up assemblages.

A simple and widely used built-up section is the *double-angle*, comprising a matched pair of angles placed back to back (Figure 7.4*f*). This member often occurs as a truss member or a bracing element.

7.3 LATERAL STABILITY OF COLUMNS

Because steel columns are quite slender—except the lower columns in very tall buildings—their lateral stability is a major concern. Designers must consider the column's unbraced length, the manner of end support (regarding any constraint against free end rotation), and any lateral bracing occurring within the column height.

Slenderness of wood and concrete columns is measured by the ratio L/d, where L is the column's unsupported length (i.e., height) and d is the least lateral dimension of the column cross section. Steel columns are never solid in cross section, so slenderness is measured by the more basic ratio L/r, where r is the *radius of gyration* of the cross section, appropriate to the buckling axis. The value of r is defined as

$$r = \sqrt{\frac{I}{A}}$$

Thus, you can compute r if you know the area of a cross section and the moment of inertia about the buckling (bending) axis. For elements such as those shown in Figure 7.3, you can find the required data for A, I, and r in AISC Manual property tables (Ref. 3). For built-up elements, such as those shown in Figure 7.4, you must independently compute values.

For a round pipe, r is the same for any centroidal axis; that is, the column has no weak axis. A square tube has two major axes, but the I and r values are the same for both axes. For shapes with major and minor axes, you must compute two I and r values.

The single angle, sometimes used as a compression strut, is special. For this shape, the x-x and y-y axes are parallel to the angle's two legs. However, the true weak axis is a diagonal axis, designated the z-z; thus this axis has the least values of I and r and so is the usual bending axis for compression buckling. *Note:* The figure at the top of Table A.2 includes this axis; the table lists values for the radius of gyration and the tangent of the angle that defines the diagonal axis.

Because some conditions affect the degree of restraint at a column's ends, relative slenderness is generally expressed as $K(L)/r$ for investigation and design. $K(L)$ is called the column's *effective length*; the modifier K accounts for various end conditions (Figure 7.5, a reproduction from the AISC Specification, provides recommended values for K).

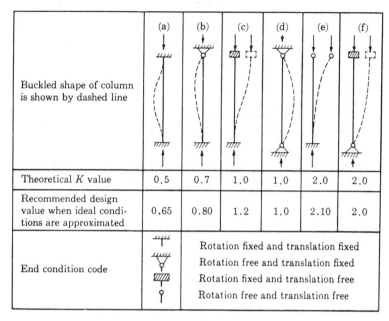

	(a)	(b)	(c)	(d)	(e)	(f)
Buckled shape of column is shown by dashed line						
Theoretical K value	0.5	0.7	1.0	1.0	2.0	2.0
Recommended design value when ideal conditions are approximated	0.65	0.80	1.2	1.0	2.10	2.0
End condition code		Rotation fixed and translation fixed				
		Rotation free and translation fixed				
		Rotation fixed and translation free				
		Rotation free and translation free				

FIGURE 7.5 Determination of effective length for buckling. Reprinted from the *Manual of Steel Construction*, 8th edition (Ref. 3) with permission of the publisher, the American Institute of Steel Construction.

The basic reference condition for determining effective length is Case d in Figure 7.5, where $K = 1.0$. In other words, no modification is made for the basic pin-ended, rotationally free, laterally restrained, buckling condition. All other cases involve some other condition at one or both ends, so K is not 1.0.

Note that Figure 7.5 lists two K values for each case. The first value is theoretical, based precisely on fully developed end conditions. The other value reflects the likelihood that reality matches theory.

I demonstrate how to use some of the values listed in Figure 7.5 later in this chapter and in other chapters. Generally, if I say nothing about end conditions for a column investigation or design, assume that $K = 1.0$.

7.4 INVESTIGATION OF COLUMNS

To determine a column's load-carrying capacity, first identify the following:

- The required compressive force (load on the column)
- The column's unbraced height
- Conditions at the column's ends that may affect its buckling; that is, establish whether K is 1.0
- What the column must do besides resist axial compression; that is, find out whether the column also must resist bending, shear, torsion, or tension
- The column type—W shape, pipe, tube, built-up section, and so on
- The grade of steel used; that is establish F_y.

Now you may investigate for one of two things:

- The load-carrying capacity of a defined column section. For example, what load can a W 14×48 carry with these conditions? How does that compare to the required load?
- The requirements for an unknown column section. Such investigation serves as preliminary design work (Section 7.6).

For both the ASD and LRFD methods, column investigation for axial compression fits predictive response equations to data from extensive laboratory tests performed over many years. When plotted, data for a single column routinely form the curves shown in Figure 7.2. In fact, Figure 7.2's curves, which reflect a safety adjustment derived from tests to ultimate failure on test columns, are used to obtain allowable stresses for the ASD method. The *original* ultimate load test curves (i.e., no safety adjustment) are the basis for investigating resistance capacity with the LRFD method.

Allowable Stress Design

You can use the formulas provided in the AISC Specification, though they are complex and cumbersome when computing by hand. However, many shortcut procedures exist, based on the fact that most columns fall into a few common groups. In fact, the AISC Manual (Ref. 3) contains many tables that list load capacities for commonly used shapes. I illustrate how to use such tables in Sections 7.5 and 7.6, as well as in Chapter 12.

For simple rolled shapes, investigation and design is easy. For built-up sections, you must use the formulas, which require a great deal of effort and patience—typically, such investigation involves a lot of trial and error.

Professional structural designers nowadays use computer software when designing columns. As a result, the following demonstrations are merely learning exercises, illuminating the process; they are not training exercises for professional design work.

AISC Manual tables list allowable axial compressive stress for various KL/r values for two F_y values: 36 and 50 ksi. Table 7.1 is a reproduction of the table for $F_y = 36$ ksi.

Load and Resistance Factor Design

The LRFD method identifies the exact mode of failure. As a result, LRFD investigations center on all conditions that may cause failure so that designers know which mode of failure predominates.

Because I limit the work in this book to relatively simple cases, I do not show the range of application of the LRFD method. In fact, the following work is done entirely with the ASD method, which sufficiently explains basic concerns and situations significant to simple structures.

7.5   USE OF TABULATED SAFE AND LIMIT LOADS

Allowable Stress Design

When investigating a column with the ASD method, you seek to establish the maximum permitted axial compressive force for service load conditions. This load limit is known as *allowable axial load* or *safe service load*. When using the analytical formulas, you establish this limit by multiplying the limiting compressive stress based on the column's KL/r by the area of the column cross section.

TABLE 7.1 Allowable Unit Stress, F_a, for Columns of A36 Steel (ksi)[a]

KL/r	F_a	KL/r	F_a	KL/r	F_a	KL/r	F_a	KL/r	F_a	KL/r	F_a	KL/r	F_a	KL/r	F_a
1	21.56	26	20.22	51	18.26	76	15.79	101	12.85	126	9.41	151	6.55	176	4.82
2	21.52	27	20.15	52	18.17	77	15.69	102	12.72	127	9.26	152	6.46	177	4.77
3	21.48	28	20.08	53	18.08	78	15.58	103	12.59	128	9.11	153	6.38	178	4.71
4	21.44	29	20.01	54	17.99	79	15.47	104	12.47	129	8.97	154	6.30	179	4.66
5	21.39	30	19.94	55	17.90	80	15.36	105	12.33	130	8.84	155	6.22	180	4.61
6	21.35	31	19.87	56	17.81	81	15.24	106	12.20	131	8.70	156	6.14	181	4.56
7	21.30	32	19.80	57	17.71	82	15.13	107	12.07	132	8.57	157	6.06	182	4.51
8	21.25	33	19.73	58	17.62	83	15.02	108	11.94	133	8.44	158	5.98	183	4.46
9	21.21	34	19.65	59	17.53	84	14.90	109	11.81	134	8.32	159	5.91	184	4.41
10	21.16	35	19.58	60	17.43	85	14.79	110	11.67	135	8.19	160	5.83	185	4.36
11	21.10	36	19.50	61	17.33	86	14.67	111	11.54	136	8.07	161	5.76	186	4.32
12	21.05	37	19.42	62	17.24	87	14.56	112	11.40	137	7.96	162	5.69	187	4.27
13	21.00	38	19.35	63	17.14	88	14.44	113	11.26	138	7.84	163	5.62	188	4.23
14	20.95	39	19.27	64	17.04	89	14.32	114	11.13	139	7.73	164	5.55	189	4.18
15	20.89	40	19.19	65	16.94	90	14.20	115	10.99	140	7.62	165	5.49	190	4.14
16	20.83	41	19.11	66	16.84	91	14.09	116	10.85	141	7.51	166	5.42	191	4.09
17	20.78	42	19.03	67	16.74	92	13.97	117	10.71	142	7.41	167	5.35	192	4.05
18	20.72	43	18.95	68	16.64	93	13.84	118	10.57	143	7.30	168	5.29	193	4.01
19	20.66	44	18.86	69	16.53	94	13.72	119	10.43	144	7.20	169	5.23	194	3.97
20	20.60	45	18.78	70	16.43	95	13.60	120	10.28	145	7.10	170	5.17	195	3.93
21	20.54	46	18.70	71	16.33	96	13.48	121	10.14	146	7.01	171	5.11	196	3.89
22	20.48	47	18.61	72	16.22	97	13.35	122	9.99	147	6.91	172	5.05	197	3.85
23	20.41	48	18.53	73	16.12	98	13.23	123	9.85	148	6.82	173	4.99	198	3.81
24	20.35	49	18.44	74	16.01	99	13.10	124	9.70	149	6.73	174	4.93	199	3.77
25	20.28	50	18.35	75	15.90	100	12.98	125	9.55	150	6.64	175	4.88	200	3.73

[a] Value of K is taken as 1.0.

Source: Adapted from data in the *Manual of Steel Construction,* 8th edition, with permission of the publisher, American Institute of Steel Construction.

Given a preselected column shape, anyone can find the column's limiting load capacity for a range of KL/r values. Long ago the publishers of the AISC Manual tabulated such load limits for future reference; today those tables yield safe service loads for almost all the rolled shapes commonly used as columns. I show how designers use data from these tables for column design in the next section.

In the following example, I illustrate the process by which single entries are derived.

Example 1.   Use a W 12 × 65 of A36 steel (F_y = 36 ksi [250 MPa]) as a column with an unbraced height of 16 ft [4.88 m]. Compute the safe service load.

Solution: From Table A.1, A = 19.1 in.^{2} [12.33 mm^2], r_x = 5.28 in. [134 mm], and r_y = 3.02 in. [76.7 mm]. No qualifying conditions are given, so you must assume that the column tends to buckle on its weak axis and K = 1.0. The critical slenderness ratio is then

$$\frac{KL}{r} = \frac{(1.0)(16 \times 12)}{3.02} = 63.6$$

To find the allowable service load stress, you can use this slenderness ratio with the AISC Specification formulas. Or you can find from Table 7.1 that a KL/r of 64 yields a stress of F_a = 17.04 ksi [117.5 MPa]. The safe service load is then

$$P = A \times F_a = 19.1 \times 17.04 = 325.5 \text{ kips } [1448 \text{ kN}]$$

For a column consisting of a built-up shape, you first must determine r. To do so, you must calculate the area of the cross section and the appropriate moments of inertia about the shape's principal axes. Then use the least value of I for the section in the formula

$$r = \sqrt{\frac{I}{A}}$$

When columns accompany other construction, such as walls or trussed bracing, the attached construction often provides some lateral bracing on one axis of the column. Without this bracing, the column may buckle with a different laterally unbraced length in other directions.

Example 2. The column in Figure 7.6 is part of a steel frame that occurs at a wall. Although the column is laterally restrained, it is free to rotate in all directions, at its top and bottom. In the direction perpendicular to the wall plane, the column has an unbraced height equal to the full height of the wall and column (L_1). However, in the wall plane beams at an intermediate height restrain the column's horizontal movement (see Figures 7.6b and c), resulting in a different unbraced height in a direction parallel to the wall (L_2). Anticipating this situation, the designer positioned the column so that its y-y axis takes advantage of the shorter unbraced length.

If the column consists of a W 12 $\times$ 58 of A36 steel, L_1 is 30 ft [9.14 m], and L_2 is 18 ft [5.49 m], what is the safe service load?

Solution. It is impossible to predetermine which axis will yield the critical value for KL/r. Using data from the preceding example, you find that the two values for KL/r are

$$x\text{-axis:} \quad \frac{KL}{r_x} = \frac{(1.0)(30 \times 12)}{5.28} = 68.2$$

$$y\text{-axis:} \quad \frac{KL}{r_y} = \frac{(1.0)(18 \times 12)}{2.51} = 86.1$$

In other words, despite the bracing, the column is still critical with respect to buckling on its weak axis—but at a shorter unbraced distance and thus with a higher allowable stress value. Using 86 for KL/r, you will find from Table 7.1 that F_a is 14.67 ksi [101 MPa]. The safe load is thus

$$P = F_a \times A = (14.67)(17.0) = 249.4 \text{ kips } [1108 \text{ kN}]$$

For the following problems, use A36 steel with F_y = 36 ksi [250 MPa] and assume that K = 1.0.

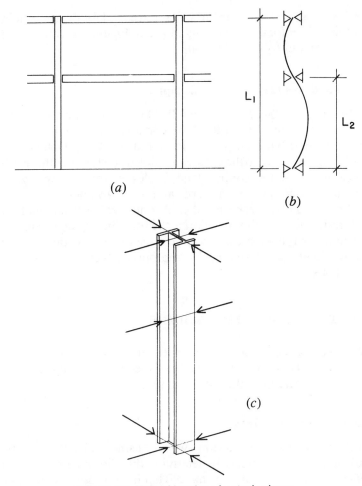

FIGURE 7.6   Biaxial bracing of a steel column.

Problem 7.5.A. Determine the safe service load for a W 10 × 88 column with an unbraced height of 15 ft [4.57 m].

Problem 7.5.B. Determine the safe service load for a W 12 × 65 column with an unbraced height of 22 ft [6.71 m].

Problem 7.5.C. Determine the safe service load for the column in Problem 7.5.B for the conditions shown in Figure 7.6, with L_1 = 15 ft [4.57 m] and L_2 = 8 ft [2.44 m].

Load and Resistance Factor Design

Tables in the AISC Manual for LRFD (Ref. 4) support shortcuts regarding columns. The LRFD method uses the same steps as the ASD method but incorporates load and resistance factors. The LRFD method cuts to the chase, and directly determines the resistance in the form of column load. Tables for rolled shapes yield resistance loads as a function of varying KL/r values.

You first must check a number of conditions to ensure that the full resistance by column action is the true limiting failure mode. Then you may compare the factored resistance load to the required design load, as determined by appropriate factored load combinations.

7.6 DESIGN OF STEEL COLUMNS

When designing steel columns without using computers, most designers rely on tabulated data. In this section I describe this process, using data from the AISC Manuals.

Allowable Stress Design

It is difficult to design a column without using load tables because it is impossible to precisely determine the allowable stress until *after* you select the column shape. That is, you cannot determine KL/r and the associated value F_a until you know r. This leads to a trial-and-error approach, which is laborious even in simple circumstances. *Note:* Trial and error is unavoidable when you need to design built-up sections, which is why their design is so tedious.

However, you can use safe load tables for single rolled shapes after determining only the required load, column height, and K factor. The AISC Manual for ASD work (Ref. 3) provides load tables for W shapes for two values of F_y, 36 ksi and 50 ksi, so you can quickly compare shapes of different yield stress.

In the following sections I offer examples that illustrate the use of AISC Manual tables. For brevity sake, I limit the examples to a single steel grade, A36—except when I cover tubular shapes, which usually are from a different steel grade.

In many cases a column's simple, axial compression capacity is the only factor affecting its design. And although columns frequently endure bending and shear (and sometimes torsion), you usually must account for the axial load capacity in combined actions. Thus the safe compression load under simple axial application conditions is vital to just about every column design.

Load and Resistance Factor Design

Professional designers who often use LRFD methods for column design undoubtedly do so with computer-aided design software—especially if the columns are part of a rigid-frame system. In this book I illustrate column design by using the ASD method, because its procedures are simpler. Remember, however, that in a full design process you must keep in mind more than just one member's structural behavior. In addition, the LRFD method is better when designing structures more complex than those treated in this book.

Note: In many situations, for simple structures, data for the different methods appear in the same general form in the AISC Manual.

7.7 SINGLE ROLLED SHAPES AS COLUMNS

The single rolled shape used most often as a column is the H shape with a nominal depth of at least 8 in. Steel rolling mills produce a wide range of such shapes; most are W shapes and some are M shapes.

AISC Manual tables list safe service loads for W and M shapes built from steel with yield strengths of 36 and 50 ksi. (Of course, the tables also include a ton of other data for the shapes in question.) Table 7.2 summarizes AISC data for shapes ranging from the W 4 × 13 to the W 14 × 730; table values are based on r for the y-y axis. Note that the table includes values for the bending factors B_x and

TABLE 7.2 Safe Service Column Loads for Selected W Shapes[a]

Shape	\multicolumn{10}{c}{Effective Length (KL) in Feet}										Bending Factor	
	8	9	10	11	12	14	16	18	20	22	B_x	B_y
M 4 × 13	48	42	35	29	24	18					0.727	2.228
W 4 × 13	52	46	39	33	28	20	16				0.701	2.016
W 5 × 16	74	69	64	58	52	40	31	24	20		0.550	1.560
M 5 × 18.9	85	78	71	64	56	42	32	25			0.576	1.768
W 5 × 19	88	82	76	70	63	48	37	29	24		0.543	1.526
W 6 × 9	33	28	23	19	16	12					0.482	2.414
W 6 × 12	44	38	31	26	22	16					0.486	2.367
W 6 × 16	62	54	46	38	32	23	18				0.465	2.155
W 6 × 15	75	71	67	62	58	48	38	30	24	20	0.456	1.424
M 6 × 20	98	92	87	81	74	61	47	37	30	25	0.453	1.510
W 6 × 20	100	95	90	85	79	67	54	42	34	28	0.438	1.331
W 6 × 25	126	120	114	107	100	85	69	54	44	36	0.440	1.308
W 8 × 24	124	118	113	107	101	88	74	59	48	39	0.339	1.258
W 8 × 28	144	138	132	125	118	103	87	69	56	46	0.340	1.244
W 8 × 31	170	165	160	154	149	137	124	110	95	80	0.332	0.985
W 8 × 35	191	186	180	174	168	155	141	125	109	91	0.330	0.972
W 8 × 40	218	212	205	199	192	177	160	143	124	104	0.330	0.959
W 8 × 48	263	256	249	241	233	215	196	176	154	131	0.326	0.940
W 8 × 58	320	312	303	293	283	263	240	216	190	162	0.329	0.934
W 8 × 67	370	360	350	339	328	304	279	251	221	190	0.326	0.921
W 10 × 33	179	173	167	161	155	142	127	112	95	78	0.277	1.055
W 10 × 39	213	206	200	193	186	170	154	136	116	97	0.273	1.018
W 10 × 45	247	240	232	224	216	199	180	160	138	115	0.271	1.000
W 10 × 49	279	273	268	262	256	242	228	213	197	180	0.264	0.770
W 10 × 54	306	300	294	288	281	267	251	235	217	199	0.263	0.767
W 10 × 60	341	335	328	321	313	297	280	262	243	222	0.264	0.765
W 10 × 68	388	381	373	365	357	339	320	299	278	255	0.264	0.758
W 10 × 77	439	431	422	413	404	384	362	339	315	289	0.263	0.751
W 10 × 88	504	495	485	475	464	442	417	392	364	335	0.263	0.744
W 10 × 100	573	562	551	540	528	503	476	446	416	383	0.263	0.735
W 10 × 112	642	631	619	606	593	565	535	503	469	433	0.261	0.726
W 12 × 40	217	210	203	196	188	172	154	135	114	94	0.227	1.073
W 12 × 45	243	235	228	220	211	193	173	152	129	106	0.227	1.065
W 12 × 50	271	263	254	246	236	216	195	171	146	121	0.227	1.058
W 12 × 53	301	295	288	282	275	260	244	227	209	189	0.221	0.813
W 12 × 58	329	322	315	308	301	285	268	249	230	209	0.218	0.794
W 12 × 65	378	373	367	361	354	341	326	311	294	277	0.217	0.656
W 12 × 72	418	412	406	399	392	377	361	344	326	308	0.217	0.651
W 12 × 79	460	453	446	439	431	415	398	379	360	339	0.217	0.648
W 12 × 87	508	501	493	485	477	459	440	420	398	376	0.217	0.645
W 12 × 96	560	552	544	535	526	506	486	464	440	416	0.215	0.635
W 12 × 106	620	611	602	593	583	561	539	514	489	462	0.215	0.633

TABLE 7.2 *Continued*

Shape	8	10	12	14	16	18	20	22	24	26	B_x	B_y
				Effective Length (KL) in Feet							Bending Factor	
W 12 × 120	702	692	660	636	611	584	555	525	493	460	0.217	0.630
W 12 × 136	795	772	747	721	693	662	630	597	561	524	0.215	0.621
W 12 × 152	891	866	839	810	778	745	710	673	633	592	0.214	0.614
W 12 × 170	998	970	940	908	873	837	798	757	714	668	0.213	0.608
W 12 × 190	1115	1084	1051	1016	978	937	894	849	802	752	0.212	0.600
W 12 × 210	1236	1202	1166	1127	1086	1042	995	946	894	840	0.212	0.594
W 12 × 230	1355	1319	1280	1238	1193	1145	1095	1041	985	927	0.211	0.589
W 12 × 252	1484	1445	1403	1358	1309	1258	1203	1146	1085	1022	0.210	0.583
W 12 × 279	1642	1600	1554	1505	1452	1396	1337	1275	1209	1141	0.208	0.573
W 12 × 305	1799	1753	1704	1651	1594	1534	1471	1404	1333	1260	0.206	0.564
W 12 × 336	1986	1937	1884	1827	1766	1701	1632	1560	1484	1404	0.205	0.558
W 14 × 43	230	215	199	181	161	140	117	96	81	69	0.201	1.115
W 14 × 48	258	242	224	204	182	159	133	110	93	79	0.201	1.102
W 14 × 53	286	268	248	226	202	177	149	123	104	88	0.201	1.091
W 14 × 61	345	330	314	297	278	258	237	214	190	165	0.194	0.833
W 14 × 68	385	369	351	332	311	289	266	241	214	186	0.194	0.826
W 14 × 74	421	403	384	363	341	317	292	265	236	206	0.195	0.820
W 14 × 82	465	446	425	402	377	351	323	293	261	227	0.196	0.823
W 14 × 90	536	524	511	497	482	466	449	432	413	394	0.185	0.531
W 14 × 99	589	575	561	546	529	512	494	475	454	433	0.185	0.527
W 14 × 109	647	633	618	601	583	564	544	523	501	478	0.185	0.523
W 14 × 120	714	699	682	663	644	623	601	578	554	528	0.186	0.523
W 14 × 132	786	768	750	730	708	686	662	637	610	583	0.186	0.521
W 14 × 145	869	851	832	812	790	767	743	718	691	663	0.184	0.489
W 14 × 159	950	931	911	889	865	840	814	786	758	727	0.184	0.485
W 14 × 176	1054	1034	1011	987	961	933	904	874	842	809	0.184	0.484
W 14 × 193	1157	1134	1110	1083	1055	1025	994	961	927	891	0.183	0.477
W 14 × 211	1263	1239	1212	1183	1153	1121	1087	1051	1014	975	0.183	0.477
W 14 × 233	1396	1370	1340	1309	1276	1241	1204	1165	1124	1081	0.183	0.472
W 14 × 257	1542	1513	1481	1447	1410	1372	1331	1289	1244	1198	0.182	0.470
W 14 × 283	1700	1668	1634	1597	1557	1515	1471	1425	1377	1326	0.181	0.465
W 14 × 311	1867	1832	1794	1754	1711	1666	1618	1568	1515	1460	0.181	0.459
W 14 × 342		2022	1985	1941	1894	1845	1793	1738	1681	1621	0.181	0.457
W 14 × 370		2181	2144	2097	2047	1995	1939	1881	1820	1756	0.180	0.452
W 14 × 398		2356	2304	2255	2202	2146	2087	2025	1961	1893	0.178	0.447
W 14 × 426		2515	2464	2411	2356	2296	2234	2169	2100	2029	0.177	0.442
W 14 × 455		2694	2644	2589	2430	2467	2401	2332	2260	2184	0.177	0.441
W 14 × 500		2952	2905	2845	2781	2714	2642	2568	2490	2409	0.175	0.434
W 14 × 550		3272	3206	3142	3073	3000	2923	2842	2758	2670	0.174	0.429
W 14 × 605		3591	3529	3459	3384	3306	3223	3136	3045	2951	0.171	0.421
W 14 × 665		3974	3892	3817	3737	3652	3563	3469	3372	3270	0.170	0.415
W 14 × 730		4355	4277	4196	4100	4019	3923	3823	3718	3609	0.168	0.408

[a] Loads in kips for shapes of steel with F_y = 36 ksi [250 MPa], based on buckling with respect to the y-axis.

Source: Adapted from data in the *Manual of Steel Construction*, 8th edition, with permission of the publisher, American Institute of Steel Construction.

B_y, which designers used when designing for combined compression and bending (see Section 7.11). Table 7.2 enables quick design selections. I demonstrate this process in Chapter 12's example building structural system designs. Such a table also serves as a check for hand calculations. For example, in Example 1 of Section 7.5, I calculated the load as 325.5 kips; from Table 7.2, the safe load for the W 12 × 65 with unbraced height of 16 ft is 326 kips.

Problem 7.7.A. Using Table 7.2, select a column section for an axial load of 148 kips [658 kN] if the unbraced height is 12 ft [3.66 m]. Use A36 steel and assume that K is 1.0.

Problem 7.7.B. Redo Problem 7.7.A, given that the load is 258 kips [1148 kN] and the unbraced height is 16 ft [4.88 m].

Problem 7.7.C. Redo Problem 7.7.A, given that the load is 355 kips [1579 kN] and the unbraced height is 20 ft [6.10 m].

7.8   STEEL PIPE COLUMNS

Round steel pipe columns occur most frequently as single-story columns that support beams (wood or steel). Pipe is available in three weight categories: standard, extra strong, and double-extra strong. A pipe's weight is independent of its inside diameter; in other words the flow in a 6 in. pipe is unaffected by the pipe's wall thickness. The dimension that varies by weight is the outside diameter. (Remember this fact when selecting pipe to fit within building construction spaces.) Pipe is designated by its nominal diameter, which approximates the inside diameter.

Table 7.3 gives safe loads for standard-weight pipe columns of A36 steel.

Example 1. Using Table 7.3, select a standard-weight steel pipe to carry a load of 41 kips [183 kN] if the unbraced height is 12 ft [3.66 m].

Solution: The table yields a value of 43 kips as the safe load for a 4 in. pipe with an unbraced height of 12 ft.

TABLE 7.3 Safe Service Column Loads for Standard Steel Pipe[a]

Nominal Dia.		12	10	8	6	5	4	3½	3
Wall Thickness		0.375	0.365	0.322	0.280	0.258	0.237	0.226	0.216
Wt./ft		49.56	40.48	28.55	18.97	14.62	10.79	9.11	7.58
F_y						36 ksi			
Effective length in ft KL with respect to radius of gyration	0	315	257	181	121	93	68	58	48
	6	303	246	171	110	83	59	48	38
	7	301	243	168	108	81	57	46	36
	8	299	241	166	106	78	54	44	34
	9	296	238	163	103	76	52	41	31
	10	293	235	161	101	73	49	38	28
	11	291	232	158	98	71	46	35	25
	12	288	229	155	95	68	43	32	22
	13	285	226	152	92	65	40	29	19
	14	282	223	149	89	61	36	25	16
	15	278	220	145	86	58	33	22	14
	16	275	216	142	82	55	29	19	12
	17	272	213	138	79	51	26	17	11
	18	268	209	135	75	47	23	15	10
	19	265	205	131	71	43	21	14	9
	20	261	201	127	67	39	19	12	
	22	254	193	119	59	32	15	10	
	24	246	185	111	51	27	13		
	25	242	180	106	47	25	12		
	26	238	176	102	43	23			
	28	229	167	93	37	20			
	30	220	158	83	32	17			
	31	216	152	78	30	16			
	32	211	148	73	29				
	34	201	137	65	25				
	36	192	127	58	23				
	37	186	120	55	21				
	38	181	115	52					
	40	171	104	47					
Properties									
Area A (in.^{2})		14.6	11.9	8.40	5.58	4.30	3.17	2.68	2.23
I (in.^{4})		279	161	72.6	28.1	15.2	7.23	4.79	3.02
r (in.)		4.38	3.67	2.94	2.25	1.88	1.51	1.34	1.16
$B \}$ Bending factor		0.333	0.398	0.500	0.657	0.789	0.987	1.12	1.29
Note: Heavy line indicates Kl/r of 200.									

[a] Loads in kips for axially loaded pipe of steel with F_y = 36 ksi [250 MPa]. Steel pipe is also available in two heavier weights.

Source: Reprinted from the *Manual of Steel Construction*, 8th edition, with permission of the publishers, American Institute of Steel Construction.

Problems 7.8.A.B.C.D. Select the minimum-size standard-weight steel pipe for an axial service load of 50 kips and the following unbraced heights: (a) 8 ft (b) 12 ft (c) 18 ft (d) 25 ft.

7.9 STRUCTURAL TUBING COLUMNS

Structural tubing, which designers use for building columns and truss members, is available in a range of sizes (nominal sizes indicate the tubes' actual outer dimensions), wall thicknesses, and steel grades. For building columns, typical sizes range upward from the 3 in. square tube. (The largest size available currently is the 16 in. square tube.)

Table 7.4, which is reproduced from the AISC Manual, yields safe column loads for two sizes of square tubes: 3 in. and 4 in.

Problem 7.9.A. Use a structural tubing column designated as TS 4 × 4 × $\frac{3}{8}$, of steel with F_y = 46 ksi [317 MPa], with an effective unbraced length of 12 ft [3.66 m]. Find the safe service axial load.

Problem 7.9.B. Select the lightest structural tubing column to carry an axial load of 64 kips [285 kN] if the effective unbraced length is 10 ft [3.05 m].

7.10 DOUBLE-ANGLES

Matched pairs of angles show up in trusses and framing braces. Usually the two angles sit back to back, separated by a gusset plate or the web of a structural tee. Compression members that are not columns are called *struts*.

The AISC Manual contains safe load tables for double-angles, assuming an average separation of $\frac{3}{8}$ in. For angles with unequal legs, two back-to-back arrangements are possible, described either as *long legs back to back*, or as *short legs back to back*. Table 7.5 is adapted from AISC tables for double-angles with long legs back to back. For each double-angle, the table lists data separately by axis, because the data depends on which axis you use to determine the effective unbraced length. If conditions relating to the unbraced

TABLE 7.4 Safe Service Column Loads for Steel Structural Tubing[a]

Nominal Size	4 × 4					3 × 3		
Thickness	½	⅜	5/16	¼	3/16	5/16	¼	3/16
Wt./ft	21.63	17.27	14.83	12.21	9.42	10.58	8.81	6.87
F_y	46 ksi							
KL = 0	176	140	120	99	76	86	71	56
2	168	134	115	95	73	80	67	53
3	162	130	112	92	71	77	64	50
4	156	126	108	89	69	73	61	48
5	150	121	104	86	67	68	57	45
6	143	115	100	83	64	63	53	42
7	135	110	95	79	61	57	49	39
8	126	103	90	75	58	51	44	35
9	117	97	84	70	55	44	38	31
10	108	89	78	65	51	37	33	27
11	98	82	72	60	47	31	27	22
12	87	74	65	55	43	26	23	19
13	75	65	58	49	39	22	19	16
14	65	57	51	43	35	19	17	14
15	57	49	44	38	30	16	15	12
16	50	43	39	33	27	14	13	11
17	44	38	34	29	24	13	11	9
18	39	34	31	26	21		10	8
19	35	31	28	24	19			
20	32	28	25	21	17			
21	29	25	23	19	16			
22	26	23	21	18	14			
23	24	21	19	16	13			
24		19	17	15	12			
25				14	11			
Properties								
A (in.²)	6.36	5.08	4.36	3.59	2.77	3.11	2.59	2.02
I (in.⁴)	12.3	10.7	9.58	8.22	6.59	3.58	3.16	2.60
r (in.)	1.39	1.45	1.48	1.51	1.54	1.07	1.10	1.13
B } Bending factor	1.04	0.949	0.910	0.874	0.840	1.30	1.23	1.17

Effective length in ft KL with respect to radius of gyration

Note: Heavy line indicates Kl/r of 200.

[a] Loads in kips for axially loaded pipe of steel with $F_y = 46$ ksi [317 MPa]. Steel tube is available in many additional sizes.

Source: Reprinted from the *Manual of Steel Construction*, 8th edition, with permission of the publisher, American Institute of Steel Construction.

length are the same for both axes, then you must use the lower value for the safe load from Table 7.5.

As with other members that lack biaxial symmetry, such as the structural tee, the slenderness of the cross section elements may affect safe load values. AISC safe load tables reflect such adjustments.

Problem 7.10.A. A double-angle compression member 8 ft [2.44 m] long consists of two A36 steel angles $4 \times 3 \times \frac{3}{8}$ in., with the long legs back to back. Determine the safe service axial compression load for the angles.

TABLE 7.5   Safe Service Axial Compression for Double-Angle Struts[a]

Size (in.)		8 × 6			6 × 4				5 × 3 1/2			5 × 3	
Thickness (in.)		3/4	1/2		5/8	1/2	3/8		1/2	3/8		1/2	3/8	5/16
Weight (lb/ft)		67.6	46.0		40.0	32.4	24.6		27.2	20.8		25.6	19.6	16.4
Area (in²)		19.9	13.5		11.7	9.50	7.22		8.00	6.09		7.50	5.72	4.80
r_x (in.)		2.53	2.56		1.90	1.91	1.93		1.58	1.60		1.59	1.61	1.61
r_y (in.)		2.48	2.44		1.67	1.64	1.62		1.49	1.46		1.25	1.23	1.22

Effective Length (KL) with Respect to Indicated Axis

X-X Axis

KL		
0	430	266
10	370	231
12	353	222
14	334	211
16	315	200
20	271	175
24	222	148
28	168	117
32	129	90
36	102	71

KL			
0	253	205	142
8	214	174	122
10	200	163	115
12	185	151	107
14	168	137	99
16	150	123	89
20	110	90	69
24	76	62	48
28	56	46	36

KL		
0	173	129
4	159	119
6	150	113
8	139	105
10	126	96
12	113	86
14	97	75
16	81	63
20	52	40

KL			
0	162	121	94
4	149	112	88
6	141	106	83
8	130	98	77
10	119	90	71
12	106	81	64
14	92	70	57
16	76	59	49
20	49	38	32

Y-Y Axis

KL		
0	430	266
10	368	229
12	351	219
14	332	207
16	311	195
20	266	169
24	216	139
28	162	106
32	124	81
36	98	64

KL			
0	253	205	142
6	222	179	125
8	207	167	117
10	190	153	108
12	171	137	97
14	151	120	86
16	129	102	74
20	85	66	49
24	59	46	34

KL		
0	173	129
4	158	118
6	148	110
8	136	101
10	122	91
12	107	79
14	90	67
16	72	53
20	46	34

KL			
0	162	121	94
4	145	108	85
6	132	99	78
8	118	88	69
10	101	75	60
12	82	61	49
14	62	46	38
16	47	35	29
20	30	22	19

Problem 7.10.B. Select a double-angle compression member for an axial compression load of 50 kips [222 kN] if the effective unbraced length is 10 ft [3.05 m].

7.11 COLUMNS AND BENDING

Steel columns frequently must sustain bending in addition to the usual axial compression. For example, Figures 7.7a to c show three of the most common such situations:

TABLE 7.5 *Continued*

	4 × 3				3 1/2 × 2 1/2				3 × 2		
	1/2	3/8	5/16		3/8	5/16	1/4		3/8	5/16	1/4
	22.2	17.0	14.4		14.4	12.2	9.8		11.8	10.0	8.2
	6.50	4.97	4.18		4.22	3.55	2.88		3.47	2.93	2.38
	1.25	1.26	1.27		1.10	1.11	1.12		0.940	0.948	0.957
	1.33	1.31	1.30		1.11	1.10	1.09		0.917	0.903	0.891
0	140	107	90,	0	91	77	60	0	75	63	51
2	134	103	86	2	86	73	57	2	70	59	48
4	126	96	81	4	80	67	53	3	67	57	46
6	115	88	74	6	71	60	48	4	63	54	44
8	102	78	66	8	61	52	41	6	55	46	38
10	88	67	57	10	50	42	34	8	44	38	31
12	71	55	47	12	37	31	26	10	32	27	23
14	54	42	36	14	27	23	19	12	22	19	16
16	41	32	27	16	21	18	15	14	16	14	12
18	33	25	22	18	16	14	12				
20	26	20	17								
0	140	107	90	0	91	77	60	0	75	63	51
2	135	103	86	2	87	73	57	2	70	59	48
4	127	97	81	4	80	67	53	3	67	56	46
6	117	89	74	6	72	60	47	4	63	53	43
8	105	80	67	8	62	52	41	6	54	45	36
10	92	70	58	10	50	42	33	8	43	36	28
12	77	58	48	12	37	31	25	10	30	25	20
14	61	45	37	14	28	23	18	12	21	17	14
16	47	35	29	16	21	17	14	14	15	13	10
18	37	27	23	18	17	14	11				
20	30	22	18								

[a] Loads in kips for axially loaded pipe of steel with F_y = 36 ksi [250 MPa]; long legs back to back; $\frac{3}{8}$ in. separation between angles.

Source: Adapted from data in the *Manual of Steel Construction*, 8th edition, with permission of the publisher, American Institute of Steel Construction.

- When loads are supported on a bracket at the column face, the eccentricity of the compression adds a bending effect (Figure 7.7a).

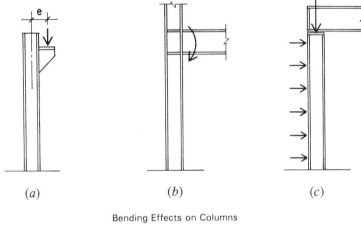

Bending Effects on Columns

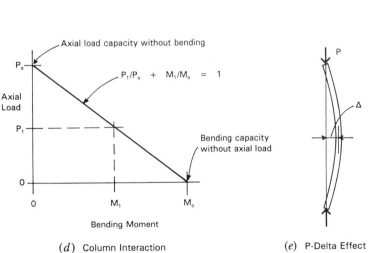

(d) Column Interaction

(e) P-Delta Effect

FIGURE 7.7 Bending in steel columns. (a) Bending induced by eccentric load. (b) Bending transferred to column in a rigid frame. (c) Combined loading condition producing axial compression and bending. (d) Classic form of compression and bending interaction. (e) Bending induced by deflection of a flexible column with an axial load.

- In a rigid frame that includes moment-resistive connections, any load on the beams induces a bending effect on the columns (Figure 7.7b).
- Columns built into exterior walls may wind up helping the wall resist wind forces (Figure 7.7c).

Adding bending to a direct compression effect results in a combined stress (or net stress) condition, in which stress is not evenly distributed across the column cross section. You may investigate the two effects separately and then add the stresses to determine the net effect, but because compression and bending are essentially different, calculating the combination of actions is a more significant investigation. Such a computation, called an *interaction* analysis, takes the form

$$\frac{P_n}{P_o} + \frac{M_n}{M_o} = 1$$

This formula describes a straight line (see Figure 7.7d), which, in elastic theory, is the classical form of the relationship between bending and compression. In real life, however, the form of the interaction is not a straight line, thanks to real-life variables, including usual columnar form and usual fabrication and construction practices.

For steel column design, formulas must take into account several major factors, including slenderness of column flanges and webs (for W shapes), ductility of the steel, and overall column slenderness (for buckling). Another potential problem is the *P*-delta effect (see Figure 7.7e), which occurs when a relatively slender column subjected to compression and bending is curved significantly by the bending effect. The deflection due to this curvature (called *delta* and designated Δ) causes an eccentricity of the compression force (i.e., a bending moment equal to the product of *P* and delta). An especially critical example: Imagine a freestanding, towerlike column with no top restraint and a load on its top. *Note:* Any bending effect on such a column can produce the *P*-delta effect.

As I mentioned above, the AISC specifications for the ASD and LRFD methods thoroughly cover the various problems introduced by combined compression and bending. However, complete treatment of this topic is well beyond the scope of this book. For more information, refer to more advanced texts (e.g., Ref. 5 and Ref. 6).

During preliminary design work, or when you want to quickly obtain a trial section for use in an extensive design investigation, you may use the following procedure, which determines an equivalent axial load that incorporates the bending effect. To do so, you use the bending factors B_x and B_y, which are given in the AISC column load tables (Table 7.2 lists these factors in its rightmost two columns).

The equivalent axial load (P') is

$$P' = P + B_x M_x + B_y M_y$$

where P' = equivalent axial compression load for design
P = actual compression load
B_x = bending factor for the column x-axis
M_x = bending moment about the column x-axis
B_y = bending factor for the column y-axis
M_y = bending moment about the column y-axis

In the following examples I illustrate how to use this approximation method.

Example 1. Use a 10 in. W shape for a column (see Figure 7.8). An axial load from above is 120 kips and a beam load at the column face is 24 kips. The column has an unbraced height of 16 ft and K is 1.0. Select a trial section for the column.

Solution: The *total* axial load is $120 + 24 = 144$ kips. Because bending occurs only about the x-axis, you use only the B_x for this case. With the exact W shape as yet undetermined, however, you must compute the equivalent load using an approximate bending factor. From Table 7.2, the range of B_x values for 10 in. shapes is quite small—0.261 to 0.277—so try using an average factor, 0.27:

$$P' = P + B_x M_x = (120 + 24) + (0.27 \times 24 \times 5)$$
$$= 144 + 32.4 = 176.4 \text{ kips}$$

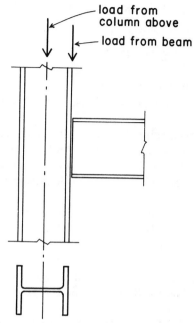

FIGURE 7.8 An eccentric loading condition with steel framing.

Use this load, which you may consider strictly an axial load, directly with Table 7.2. However, the table values are based on the y-axis, and the bending in this case is with respect to the x-axis. Therefore, a more accurate use of this P' value is to compare it to the column capacity based on the slenderness ratio KL/r_x.

For example, if you choose a shape from Table 7.2 based on 176.4, you wind up with a W 10 × 45 with a capacity of 180 kips. But if you ignore for the moment the bending action, you can choose a shape for a load of 144 kips. Then you wind up with a W 10 × 39 with a capacity of 154 kips, which, based on the weak axis, is the minimum column. From Table A.1, you find for this shape that $A = 11.5$ in.^{2} and $r_x = 4.27$ in. Now compute

$$\frac{KL}{r_x} = \frac{(16 \times 12)}{4.27} = 44.96, \text{ say } 45$$

From Table 7.1, F_a = 18.78 ksi, so the axial load allowable, based on the x-axis, is

$$P_x = F_a \, A = (18.78)\,(11.5) = 216 \text{ kips}$$

Clearly, the W 10 × 39 is adequate.

Although this process is laborious, using AISC formulas is even more laborious.

When bending occurs about both axes, as in full three-dimensional rigid frames, you must use all three parts of the approximation formula. In the end, you can choose a shape directly from Table 7.2. In the following example I demonstrate this process.

Example 2. Use a 12 in. W shape for a column that sustains the following: axial load of 60 kips, M_x = 40 kip-ft, M_y = 32 kip-ft. Select a column for an unbraced height of 12 ft.

Solution: From Table 7.2, for 12 in. shapes, you can determine approximate values for B_x (0.215) and B_y (0.63). Thus

$$P' = P + B_x M_x + B_y M_y$$
$$= 60 + \{0.215 \times (40 \times 12)\} + \{0.063 \times (32 \times 12)\}$$
$$= 60 + 103 + 242 = 405 \text{ kips}$$

From Table 7.2, the lightest 12-ft-tall shape is a W 12 × 79, with an allowable load of 431 kips and true bending factors of B_x = 0.217 and B_y = 0.648. Because these factors slightly exceed those assumed, find a new value for P' to verify this choice of shape. Thus

$$P' = 60 + \{0.217 \times (40 \times 12)\} + \{0.648 \times (32 \times 12)\}$$
$$= 413 \text{ kips}$$

The choice is still valid.

As I mentioned at the beginning of this section, columns in rigid-frame bents often face bending. In fact, in an all-steel frame, beams frame columns on both sides—or, indeed, *all* sides when the columns are interior columns. I further discuss such a frame in Chapter 8.

Problem 7.11.A. Use a 12 in. W shape for a column to support a beam as shown in Figure 7.8. Select a trial column size given the following: column axial load from above = 200 kips, beam reaction = 30 kips, unbraced column height is 14 ft.

Problem 7.11.B. Use a 14 in. W shape for a column that sustains bending on both axes. Select a trial column section given the followiing: total axial load = 160 kips, M_x = 65 kip-ft, M_y = 45 kip-ft, unbraced column height is 16 ft.

7.12 COLUMN FRAMING AND CONNECTIONS

When developing connection details for columns, you must consider the columns' shape and size; the other framing's shape, size, and orientation; and the joints' particular structural functions. Some common forms of simple connections for light frames are shown in Figure 7.9. Connections are usually attached by welding or bolting (with high-strength steel bolts or anchor bolts embedded in concrete or masonry).

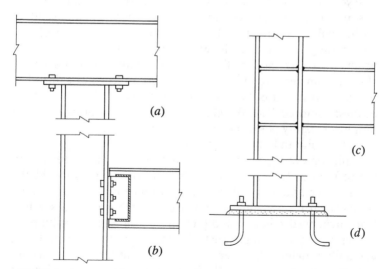

FIGURE 7.9 Typical fabrication details for steel columns in lightly loaded frames.

When beams sit on top of a column (Figure 7.9a), designers usually weld a bearing plate on top of the column and bolt the beam's bottom flanges to the plate. For all connections, you must consider what parts of the connection are completed in the fabrication shop and what is done at the job site (called the *field*). The plate usually is attached to the column in the shop (where welding is preferred), while the beam usually is attached in the field (where bolting is preferred). In this joint the plate serves no structural function; the beam can bear directly on top of the column. However, the plate makes field assembly of the frame easier. In addition, the plate may help spread the bearing stress more fully over the column cross section.

In many situations beams must frame into a column's side. If all you need is a simple transfer of vertical force, a common solution is to use a pair of steel angles to connect the beam web to the column face (see Figure 7.9b). You can use a variation of this connection to connect a beam to the column web when framing intersects the column differently, as long as the outspread legs of the angles fit between the column flanges. In general, such a connection requires at least a 10 in. W shape column, which partially explains why 10, 12, and 14 in. W shapes are popular for columns.

If you must transfer bending moment between a beam's end and its supporting column, a common solution is to weld the cut ends of the beam flanges directly to the column face (see Figure 7.9c). Since the bending must be developed in both column flanges but the beam grabs only one directly, designers use filler plates for a more effective transfer. Of course, you also must connect the beam web to the column because the beam web actually carries most of the beam shear force. *Note:* Widely used for years, this form of connection has recently performed poorly in earthquakes.

At the column bottom, where bearing often is on top of a concrete pier or footing, designers try to reduce the bearing pressure on the soft concrete. Given 20 ksi or more of compression in the column steel, but only 1000 psi of resistance in the concrete, the contact bearing area must be spread out. For this reason, as well as a practical need to hold the column in place, a common solution is to attach a steel bearing plate to the column in the shop and then add a filler material between the smooth plate and the rough concrete surface (see Figure 7.9d). This form of connection is adequate for lightly loaded columns, but designers modify it to transfer very high column loads, develop uplift or bending moment resistance, and so on. Still, the simple joint is the most common form.

For tall steel frames, you usually must splice columns—there's a limit to how long a single piece can be. Besides, shorter pieces are easier to handle. For example, if you pick up a very small but very long W shape (with its significantly weak y-axis), its own weight can cause it to bend permanently. On the other hand, *all* joints are relatively expensive, so the frame with the fewest joints is likely to be the least expensive. Therefore, most designers use as long a single piece possible—at least two stories in most multistory construction.

Figure 7.10 shows some ways to splice stacked pieces. In these joints, the upper column part bears directly on the end of the lower column part to transfer vertical load; the bolted plates keep the parts from slipping sideways. A problem: matching the upper and lower parts permits a contact bearing. For example, the joint in Figure 7.10*a* works only if the two column parts have similar overall depths (i.e., out-to-out dimension for the flanges) and if the web of the upper part is thinner than the web of the lower part.

In Figure 7.10*b*, the *inside* dimension between flanges is the same so the bearing contact still works, though a gap exists between the splice plates and the face of the flanges in the upper part. (You can fill this gap by using filler plates.) This case is common because most W shapes of similar nominal depth have similar inside dimensions.

If column flanges are totally unmatched, you can use a very thick bearing plate—one that covers the whole top of the lower column—to achieve a bearing connection.

You can develop column splice joints as completely welded ones, especially when the structure is a rigid frame.

I cover various considerations for development of both bolted and welded connections for steel framing in Chapter 11. For more detail, refer to AISC publications.

7.13 COLUMN BASES

Force transfer to supporting materials at the base of a column frequently involves only direct bearing stress and so requires only a base plate (see Figure 7.9*d*). For very light column loads, the plate

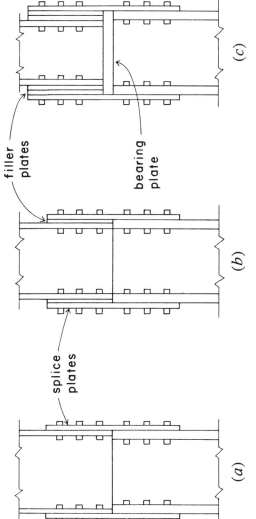

FIGURE 7.10 Typical bolted splices for multistory columns.

area required is often less than the column's footprint (i.e., product of the shape depth times the flange width); however, you must use a plate whose area is larger than the column's footprint if only to make it easier to place anchor bolts and weld the plate to the column.

Use the following procedure to determine the required dimensions for a column base plate, assuming that the required size is significantly larger than the column footprint. The required base plate area is

$$A_1 = \frac{P}{F_p}$$

where A_1 = plan area of the base plate
P = compression load from the column
F_p = allowable bearing stress on the concrete, which is based on the concrete design strength, f_c' .

If the plate covers the entire area of the supporting concrete (such a situation is rare but serves as a reference point), bearing is limited to $0.3f_c'$. If the area of the supporting member is larger, you may increase bearing by a factor equal to $\sqrt{A_2/A_1}$ but not larger than 2, where A_2 is the area of the supporting concrete.

Figure 7.11 shows, for a W shape column, how to determine plate thickness due to bending. Once you find A_1 , establish B and N so that the projections m and n are approximately equal. *Note:* You must relate the dimensions to the locations of anchor bolts and to the details regarding attaching the plate to the column.

The required plate thickness, based on bending stress, is determined by

$$t = \sqrt{\frac{3f_p m^2}{F_b}} \text{ or } t = \sqrt{\frac{3f_p n^2}{F_b}}$$

where m and n = the dimensions shown in Figure 7.11
t = plate thickness
f_p = actual bearing pressure, P/A_1
F_b = allowable bending stress in the plate, $0.75F_y$

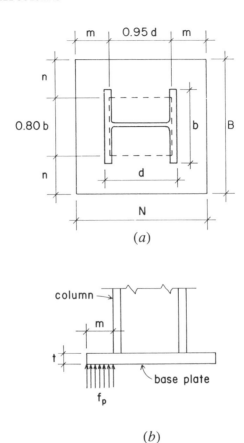

(a)

(b)

FIGURE 7.11 Reference dimensions for column base plates.

In the following example I illustrate this process for a column with a relatively light load.

Example 1. Design a base plate of A36 steel for a W 10 × 49 column with a load of 250 kips. The column bears on a concrete footing with $f_c' = 3$ ksi.

Solution: Assume that the footing plan area is considerably larger than the base plate, so use the maximum increase factor, 2, for F_p. Thus $F_p = 2(0.3 \ f_c') = (0.6)(3000) = 1800$ psi, or 1.8 ksi. Then

$$A_1 = \frac{P}{F_p} = \frac{250}{1.8} = 138.9 \text{ in.}^2$$

To approximate the dimensions, find

$$\sqrt{138.9} = 11.8 \text{ in.}$$

The layout shown in Figure 7.12 is possible: It uses full inch dimensions, allows for welds and anchor bolts, and results in a reasonably equal m and n. Because m (2.26) is slightly larger than n (2.0), the plate thickness is

$$t = \sqrt{\frac{3f_p m^2}{F_b}}$$

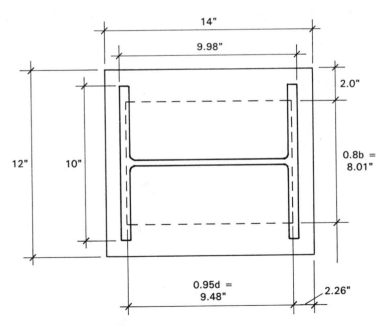

FIGURE 7.12 Reference for Example 1.

for which the actual soil pressure is

$$f_p = \frac{P}{A_1} = \frac{250}{(12 \times 14)} = 1.49 \text{ ksi}$$

and

$$t = \sqrt{\frac{3(1.49)(2.26)^2}{0.75(36)}} = \sqrt{0.846} = 0.9195 \text{ in.}$$

Plates are usually specified in thickness increments of $\frac{1}{8}$ in., so a practical specified thickness is 1.0 in.

Problem 7.13.A. Design a column base plate for a W 8 × 31 column that carries a load of 178 kips and rests on a concrete pier with $f'_c = 2500$ psi. The pier's side dimensions almost equal the base plate's plan size.

Problem 7.13.B. Redesign the base plate for the column in Problem 7.13.A, given that bearing is on an 8-ft-square footing of the same grade concrete.

8

FRAMED BENTS

Frames consisting of vertical columns and horizontal-spanning members are often arranged as simple post-and-beam systems whose columns function as simple, axially loaded compression members and whose horizontal members function as simple beams. In some cases, however, the interactions between frame members may be more complex. For example, in the rigid frame members are connected for moment transfer, and in the braced frame diagonal members produce truss actions. In this chapter I consider the behavior and design of rigid and braced frames.

8.1 DEVELOPMENT OF BENTS

A *bent* is a planar frame formed to resist lateral loads, such as those produced by wind or earthquakes. The simple frame in Figure 8.1*a* consists of three members connected by pinned joints with pinned joints also at the bottom of the columns. In theory, this frame is stable under vertical load only if both the load and the frame are perfectly symmetrical. However, any lateral (in this case, horizontal)

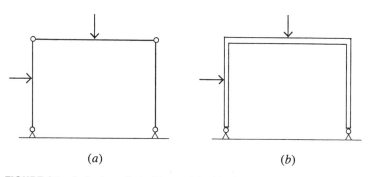

(a)   (b)

FIGURE 8.1 A single-unit steel frame (a) with all pinned connections (typical post-and-beam construction) (b) with beam-to-column moment-resistive connections, producing a rigid frame.

load or slightly unbalanced vertical load topples the frame. One means for restoring stability is to connect the columns' tops to the beam's ends with moment-resistive connections (see Figure 8.1*b*). Then the frame deformation under vertical loading is as shown in Figure 8.2*a*, with moments developed in the beam and columns. The frame deformation under lateral load is as shown in Figure 8.2*b*.

When you transform a frame into a rigid-frame bent, primarily to achieve lateral stability, the frame's response to vertical load is unavoidably altered. In the frame shown in Figure 8.1*a*, the columns (stable or not) are subject to only vertical axial compression under vertical loading on the beam; when connected to the beam to produce rigid frame action, they also are subject to bending. Given the combination of vertical and lateral loads, the bent functions as shown in Figure 8.2*c*.

8.2 MULTIUNIT RIGID FRAMES

Single-unit rigid frames show up in single-space, one-story buildings, while multiunit rigid-frame bents are the norm in multistory buildings with multiple horizontal bays.

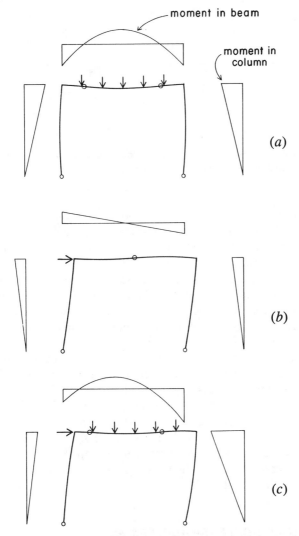

moment in beam

moment in
column

(a)

(b)

(c)

FIGURE 8.2 Actions of the single-unit rigid frame (a) under gravity load (b) under lateral load (c) under combined gravity and lateral loading.

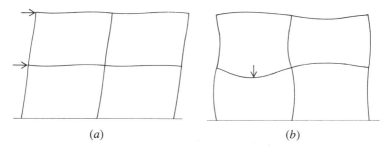

(a)

(b)

FIGURE 8.3 Actions of a multistory, multiple-span rigid frame (a) under lateral loads (b) with gravity load on a single beam.

Figure 8.3*a* shows the response of a two-story, two-bay rigid frame to lateral loading: all the frame's members are bent, which means they all help resist the loading. Even if only a single member is loaded (see Figure 8.3*b*), all the frame's members respond.

The term *rigid* actually applies to a frame's connections; in other words, the joint resists deformation sufficiently to prevent any significant rotation of one connected member with respect to the other. *Note:* In steel frames most moment-resistive connections are the result of welding.

"Rigid" does not describe the frame's nature regarding lateral resistance. Other means for bracing frames against lateral loads (by shear panels or trussing) typically produce stiffer structures.

In general, rigid frames are statically indeterminate, so their investigation and design is beyond the scope of this book. I discussed designing for combined compression and bending in Chapter 7, and I detail moment-resistive connections in Chapter 11. I illustrate the approximate design of rigid bents in the design examples in Chapter 12.

8.3 THREE-DIMENSIONAL FRAMES

Most building structures with column-and-beam frames have ordered layouts: typically, columns are evenly spaced in regularly spaced rows; in turn, the beams are also regularly spaced. In such sysems vertical planes of columns and beams define series of two-dimensional bents (see Figure 8.4). Such a bent does not exist as a total, freestanding entity ,but as a subset of the structure. However, you may investigate how an individual bent responds to gravity and lateral loads.

Multiunit frames are typically also three-dimensional with regard to the total building framework. Rigid frame actions may be three-dimensional or may be limited to the actions of selected planar (i.e., two-dimensional) bents. With sitecast concrete frames, three-dimensional frame action is natural and generally unavoidable. With wood frames and steel frames, because the normal framework is a simple post-and-beam system, you must add moment-resistive joints between beams and columns to produce rigid-frame actions.

For any rigid frame, you must acount for the action of the whole frame—that is, the collective *interactions* of all the columns and beams. Unfortunately, such interactions are highly statically indeterminate. For large frames, investigating this behavior, which includes possible variations in loading, is a formidable computational problem that precedes any design efforts.

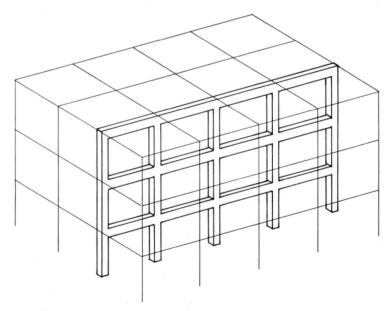

FIGURE 8.4 Planar (i.e., two-dimensional) column-beam bent as a subset in a three-dimensional framing system.

Designers often use the vertical rigid-frame bent to brace for lateral forces due to wind and earthquakes. Once such a bent is constituted as a rigid (i.e., moment-resisting) frame, however, its continuous, indeterminate responses occur for all loadings. In turn, designers must consider its response to gravity load, lateral load, and the combination of gravity and lateral loadings.

The complete investigation and design of multiunit rigid frames is well beyond the scope of this book. For some general considerations regarding the design of multistory frames, see the discussion for example Building Five in Chapter 12.

8.4 MIXED FRAME AND WALL SYSTEMS

Most buildings include both framed systems and walls. Walls vary in structural potential. For instance, metal and glass skins are typically not structural components, even though they must have some structural character to resist gravity and wind effects. Meanwhile, walls of cast concrete or concrete masonry construction frequently serve as structural parts.

When walls are structural components, you must analyze the relationships between the walls and the framed structure. In this section I outline this analysis.

Coexisting, Independent Elements

Frames and walls act independently for some functions, and they interact for other purposes. For low-rise buildings, walls often brace the building for lateral forces, even when a complete gravity-load-carrying frame structure exists (e.g., a light wood frame construction with plywood shear walls, a concrete frame structure with cast concrete or concrete masonry walls).

How you attach walls and frames depends on the actions desired; remember, walls typically are very stiff, while frames often experience significant deformation due to bending in the frame members. If you expect walls and frames to interact, you must connect them rigidly to achieve the necessary load transfers. However, if you want them to act independently, you may need to develop special attachments that allow load transfer sometimes, independent movements at other times, and total separation at still other times.

You may design a frame for gravity load resistance only, with lateral load resistance developed by walls acting as shear walls (see Figure 8.5). Such a design usually requires some frame elements to function as collectors, stiffeners, shear wall end members, or chords for horizontal diaphragms. If you intend to use the walls strictly for lateral bracing, you must carefully attach wall tops to overhead beams: you must permit deflection of beams to occur without transferring loads to the walls.

Load Sharing

Walls that are rigidly attached to columns usually provide continuous lateral bracing in the plane of the wall. This frame design, which permits you to design the column only for the relative slenderness perpendicular to the wall, is often useful for a wood or steel column (e.g., a wood stud 2 × 4 and a steel W shape with narrow flanges) and occasionally significant for a concrete or masonry column whose cross section is not square or round.

In some buildings you may use walls and frames for lateral load resistance at different locations or in different directions. Figure 8.6 shows four such situations:

- In Figure 8.6a a shear wall and a parallel frame at different ends of the building share the wind load from one direction equally.

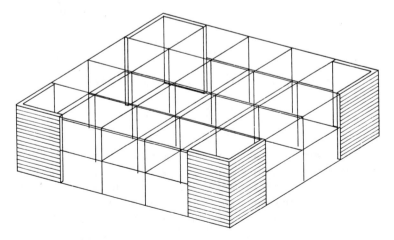

FIGURE 8.5 Framed structure braced by shear walls.

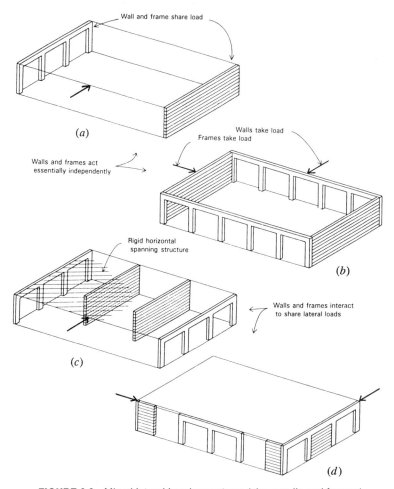

FIGURE 8.6   Mixed lateral bracing systems (shear walls and frames).

- In Figure 8.6b walls and frames are used for lateral loads in perpendicular directions. Although some load is distributed among the walls in one direction and the frames in the other direction, the walls and frames essentially do not interact unless the building faces some significant torsion.

- Figures 8.6c and d show situations in which walls and frames do interact to share the total load from a single direction. If the horizontal structure is reasonably stiff in its own plane, the load sharing is based on the relative stiffnesses of the vertical elements. (*Relative stiffness* refers to resistance to deflection under lateral force.)

Dual Systems

In a *dual system* for lateral bracing, a shear wall system deliberately shares loads with a frame system. In Figure 8.6 the systems shown at a and b are not dual systems, but those shown at c and d may be. The dual system improves structural performance, but the construction must be carefully designed and detailed to ensure that interactions and deformations do not cause excessive damage to the general construction. I discuss some special problems in the next section.

8.5 SPECIAL PROBLEMS OF STEEL RIGID-FRAME BENTS

The following problems may crop up if you use rigid-frame bents within a three-dimensional framing system:

Lack of Symmetry of Columns. A common shape used for steel columns is the W shape, which has a strong axis (the x-axis) and a weak axis (the y-axis). While planning the layout of the framework, you must decide how to orient the columns. For rigid-frame bent actions, the beams of the bent should frame into the column flanges, not the column web. However, other considerations are important, including whether to make double use of a column, as a member of perpendicular bents.

Pinned Versus Rigid Connections. As I mention elsewhere, the connections between beams and columns in a basic steel structure do not develop significant moment resistance. Therefore rigid-frame bents require special connections. However, this need enables designers to control which frame members participate in the bent action. In fact, in most buildings that use steel rigid bents, only selected members function in bent actions; the rest are merely along for the ride.

Shop Versus Field Work. Erecting steel frames is most economical when most of the fabrication is done in factory conditions (the *shop*) and connections done at the job site (the *field*) are simple and few. In addition, on-site work not only may delay the erection, but it usually is not as good as factory work. However, producing large rigid bents usually requires a great deal of field work.

General Cost. Rigid-frame bents are popular but expensive. They require extra design effort, special connections, and extensive field welding. In addition, columns that are designed for combined compression and bending must be bigger—and steel isn't cheap.

Proportionate Stiffness of Individual Bent Members. The behavior of a bent and the forces in a bent's members are strongly affected by the members' proportionate stiffness. And if story heights vary and beam spans vary, you may encounter some very complex and unusual behavior.

Of particular concern is the relative stiffnesses of all the columns in a single story of the bent. In many cases lateral shear is distributed on this basis—that is, the stiffer columns carry more lateral force.

Another concern: the relative stiffnesses of columns compared to beams. Most bent analyses assume the column stiffness to be more-or-less equal to the beam stiffness, producing the classic form of lateral deformation shown in Figure 8.3a; individual bent members take an S-shaped, inflected form. However, if the columns are much stiffer than the beams, the deformation may take the form shown in Figure 8.7a, with virtually no inflection in the columns and an excessive deformation in upper beams. This case is prevalent in tall frames because gravity loads require large columns in the lower stories.

Conversely, if the beams are much stiffer than columns, the deformation may take the form shown in Figure 8.7b, with columns behaving as if fully fixed at their ends. This situation is common in structures with deep spandrel beams and relatively small columns.

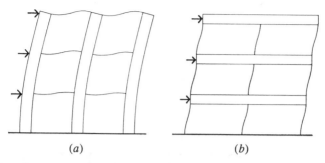

(a) (b)

FIGURE 8.7 Deformation under lateral loading of rigid frames with members of disproportionate stiffness: (a) stiff columns and flexible beams (b) stiff beams and flexible columns.

The Captive Frame. Another problem is that of the partially re-strained column or beam, where inserted construction alters the form of bent deformation. An example is the *captive column.* In Figure 8.8, for example, a partial height wall sits between columns. If this wall has sufficient stiffness and strength and is tightly wedged between columns, the col-umns' laterally unbraced height is drastically altered. As a result, the shear and bending in the columns are considerably different than what was calculated for the free columns. In addition, the distribution of forces in the bent containing the captive columns may change. Finally, if the bent is signifi-cantly stiffer, its share of the load is higher than that of other parallel bents.

This issue should concern not only the structural designer, but also whoever does the construction detailing for the wall construction.

Note: The captive frame is especially problematic for con-crete frames affected by seismic forces.

8.6 TRUSSED FRAMES

The term *braced frame* describes a frame with diagonal members that produce some truss action. In other words, adding some di-

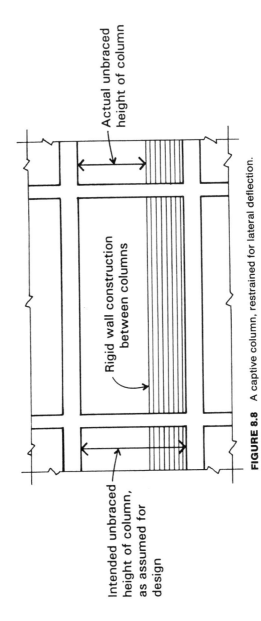

FIGURE 8.8 A captive column, restrained for lateral deflection.

Actual unbraced height of column

Rigid wall construction between columns

Intended unbraced height of column, as assumed for design

agonals to typical column-and-beam rectilinear layouts produces trussed bents that function as vertically cantilevered trusses, providing lateral bracing.

Figure 8.9 shows two forms of such arrangements:

- In Figure 8.9*a* the addition of single diagonals produces a simple truss that is statically determinate. A disadvantage of this form is that the diagonals must function for lateral force from both directions. Thus they are sometimes in compression and sometimes in tension. When functioning in compression, the very long diagonals must be quite heavy to prevent buckling on their unbraced lengths.

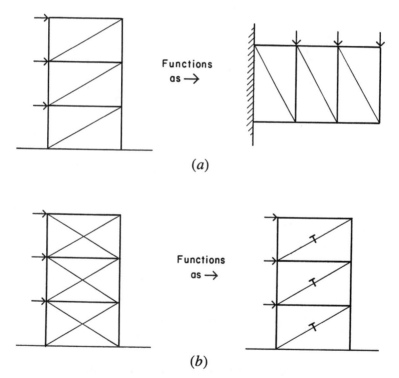

FIGURE 8.9   Frames with concentric bracing: (a) basic form of the cantilever action under lateral loading (b) assumed limit condition for the frame with light X-bracing.

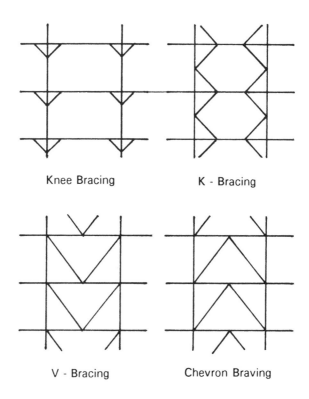

Knee Bracing   K - Bracing

V - Bracing   Chevron Braving

Semi - Concentric Bracing, Beam - to - Column

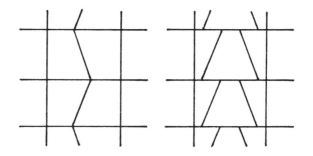

Eccentric Bracing, Beam - to - Beam

FIGURE 8.10 Common forms of eccentric bracing.

- Another popular trussed bracing is *X-bracing* (see Figure 8.9*b*). Designers who use this form sometimes use very slender tension members (perhaps long round rods), with a single rod functioning in tension for each direction of loading. If designers assume that the other rod buckles slightly in compression and becomes negligible in resistance, they may analyze the X-braced frame by simple statics, even though it is theoretically indeterminate with all members working. *Note:* Light X-bracing can improve a structure's resistance to wind forces, but not seismic forces (unless the compression diagonals are quite stiff).

Both single and double diagonals use up space, making it hard to place windows or doors within the bay. Although both knee bracing and K-bracing leave more open space in the bay (see Figure 8.10), they cause bending in the columns and beams and involve a form of combined truss and rigid-frame action. The V-brace and inverted V, or chevron brace, develop bending only in the beams.

All the aforementioned bracing forms successfully brace against wind. However, the dynamic jerking and rapid reversals in direction of seismic forces rule out all but the V-brace and chevron brace. Recently designers began using *eccentric bracing*, whose sloped bracing members connect only at points within the beam spans, for seismic bracing. The eccentric braces act as the weak links, failing in inelastic buckling with some degree of yielding; as a result, this system is highly suited to dynamic loading conditions.

9

MISCELLANEOUS
STEEL COMPONENTS
AND SYSTEMS

Most steel structures use rolled structural products and formed sheet steel units for common structural systems. In this chapter I briefly discuss some other uses of steel for building structures.

9.1 MANUFACTURED SYSTEMS

Some manufacturers produce complete buildings. The "package" building may come totally assembled (as with a mobile home) or in a kit of components (some assembly required!).

Although much of this construction lacks architectural grace (the army Quonset hut, Butler Brothers corn cribs), a variety of buildings use these systems, including housing and schools, as well as utilitarian industrial and agricultural buildings. Designers may choose prefab construction because cost is known. Many such systems employ steel because it is light, strong, dependable, and non-combustible.

Often, manufacturers patent components of these systems in an attempt to make its system unique. However, because most elements used for steel structures are standardized, the form of the end product is predictable. Designing steel structure is largely a matter of assembling a system from predetermined parts.

9.2 COMPOSITE STRUCTURAL ELEMENTS

The term *composite* usually refers to structural elements that comprise two or more materials with significantly different stress-strain characters. The most common composite structural element is reinforced concrete.

Steel-Concrete Beams

A common construction form for floors with steel framing consists of sitecast concrete slabs atop steel beams (see Figure 6.13b). If the concrete slab is mechanically anchored by some means to the beams' top flanges, it contributes to the development of compression force for a combined resisting moment. Figure 9.1 shows a

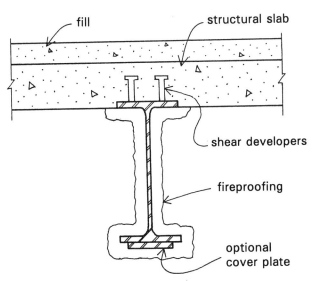

FIGURE 9.1 Composite construction: steel beams and a sitecast concrete deck.

typical anchor form: called *shear connectors, shear developers,* or *studs,* the anchors are welded to the top of the beam flange and engage the sitecast concrete. Anchors not only enhance a steel framing's strength, they also reduce deflections, a more significant gain in many cases. The AISC Manual contains a sample design of such construction (refer to the "Composite Design for Building Construction" section). The manual also has tables of transformed section modulus values, measuring the beam's increased moment resistance.

Flitched Beams

Sometimes in construction with solid timber beams, beam deflection is critical. More specific, long-term sag is inevitable when you build beams from green-wood timber. To control such sag, designers often combine a timber beam with steel plates—the *flitched beam.* Sometimes designers may sandwich the wood member between two plates (Figure 9.2*b*), but more often they place a single plate between two wood members (Figure 9.2*a*). Flitched beam components are held together with through bolts so that the elements act together as a single unit.

When computing flexural stresses, assume that the two materials deform equally. Let

Δ_1 and Δ_2 = deformations per unit length (strain) of the outermost fibers of the two materials

f_1 and f_2 = unit bending stresses in the outermost fibers of the two materials

E_1 and E_2 = moduli of elasticity of the two materials

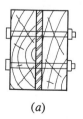 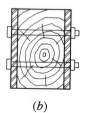

(*a*)   (*b*)

FIGURE 9.2 The flitched beam.

By definition, the elastic modulus of a material is equal to the unit stress divided by the unit strain, so

$$E_1 = \frac{f_1}{\Delta_1}$$

and

$$E_2 = \frac{f_2}{\Delta_2}$$

Transposing,

$$\Delta_1 = \frac{f_1}{E_1}$$

and

$$\Delta_2 = \frac{f_2}{E_2}$$

Since the two deformations must be equal,

$$\frac{f_1}{E_1} = \frac{f_2}{E_2}$$

or

$$f_2 = f_1 \times \frac{E_2}{E_1}$$

This simple equation, which defines the relationship between the stresses in the two materials of a composite beam, serves as the basis for investigating a flitched beam, as I demonstrate in the following example.

Example 1. A flitched beam consists of two 2 × 12 planks of Douglas fir, No. 1 grade, and a 0.5 × 11.25 in. [13 × 285 mm] A36 steel plate. Compute the allowable uniformly distributed load this beam carries on a simple span of 14 ft [4.2 m]. For the steel, $E = 29{,}000{,}000$ psi [200 GPa] and the maximum allowable bending stress is 22 ksi [150 MPa]. For the wood, $E = 1{,}800{,}000$ psi [12.4 GPa] and the maximum bending stress is 1500 psi [10.3 MPa]. For the 2 × 12, $S = 31.6$ in.^{3} [518 × 10^3 mm^3].

Solution: Assume that the steel stress is critical and find the corresponding wood stress:

$$f_w = f_s \times \frac{E_w}{E_s}$$

$$= \frac{(22{,}000)\,(1{,}800{,}000)}{29{,}000{,}000}$$

$$= 1366 \text{ psi} \ [9.3 \text{ MPa}]$$

You know that your assumption was correct because 1366 psi is lower than the wood's maximum allowable stress. In other words, a stress higher than 1366 psi exceeds the steel's limit but not the wood's.

Now find the share of the load carried by the wood members. The moment on the simple beam due to this load (W_w) is

$$M = \frac{W_w L}{8} = \frac{W_w \times 14 \times 12}{8} = 21 \ W_w$$

Using S of 31.6 in.^{3} for the 2 × 12,

$$M = f_w \times S = 1366(2 \times 31.6) = 21 \ W_w$$

$$W_w = 4111 \text{ lb } [18.35 \text{ kN}]$$

For the steel plate,

$$S = \frac{bd^2}{6} = \frac{(0.5\,(11.25)^2}{6}$$

$$= 10.55 \text{ in.}^3 \; [176 \times 10^3 \text{ mm}^3]$$

Then

$$M = 21\,W_s = f_s \times S = (22,000)\,(10.55)$$

$$W_s = 11,052 \text{ lb } [50.29\text{kN}]$$

and the total capacity of the combined section is

$$W = W_w + W_s = 4111 + 11,052 = 15,163 \text{ lb } [68.64 \text{ kN}]$$

Although the wood's load-carrying capacity is actually reduced in the flitched beam, the resulting total capacity is substantially greater. This significant increase in strength, accompanied by only a small increase in size explains the flitched beam's popularity.

Problem 9.2.A. A flitched beam consists of a single 10 × 14 of Douglas fir, Select Structural grade, and two A36 steel plates, each 0.5 × 13.5 in. [13 × 343 mm]. Compute the single concentrated load this beam can safely support at the center of a 16 ft [4.8 m] simple span. Neglect the weight of the beam. For the steel, the limiting bending stress is 22 ksi. For the wood, $E = 1,600,000$ psi [11.03 GPa] and the allowable bending stress is 1600 psi [11.03 MPa]. For the 10 × 14, $S = 228.6$ in.^{3} [3.75 × 10^6 mm^3].

Formed Sheet Steel Deck Plus Concrete Fill

When designers combine formed sheet steel decks with concrete fill of a structural grade (a normal floor construction), the steel and concrete elements act in one of three ways:

- The concrete functions only as an inert fill material; the steel deck is the structural support.

- The steel deck functions only to support the wet, freshly cast concrete; when hardened, the concrete serves as a spanning reinforced concrete slab deck.
- After the concrete dries, it interacts with the concrete (i.e., a composite action). In fact, the steel deck resists positive bending moments on the reinforced slab; that is, the steel deck resists midspan moments that cause tension in the bottom of the slab. Such steel decks have lugs or indentations, similar to the raised ridges on ordinary reinforcing bars, so that it may better engage the cast concrete.

9.3 TENSION ELEMENTS AND SYSTEMS

Some steel tension elements are simple—for example, a hanger or tie rod—and some are complex—for example, cable networks or restraining cables for tents or pneumatic structures. In this section I outline the actions of some simple steel tension elements.

Axially Loaded Tension Elements

The simplest case of tension stress occurs when a linear element is subjected to tension and the tension force is aligned on an axis that coincides with the centroid of the element's cross section. For such an element the tension stress (assumed to be distributed evenly on the cross section) is expressed as

$$f = \frac{T}{A}$$

The unit deformation (strain) for stress in the elastic range is

$$\varepsilon = \frac{f}{E}$$

For a member with length L, the total change in length (stretch) is

$$e = \varepsilon L = \frac{f}{E}L = \frac{TL}{AE}$$

In turn,

$$T = fA = \frac{AEe}{L}$$

When a tension member is short (e.g., a short hanger or a truss member), designers base the usable tension capacity on the limiting tension stress. However, for long members, elongation may be critical, limiting the capacity well below the usual safe stress limit.

Example 1. An arch spans 100 ft [30 m] and is tied at its spring points by a round steel rod with a 1 in. [25 mm] diameter. Find the limit for the tension force in the rod if stress is limited to 22 ksi [150 MPa] and total elongation is limited to 1.0 in. [25 mm].

Solution: The cross-sectional area of the rod is

$$A = \pi R^2 = 3.14\,(0.5)^2$$

$$= 0.785 \text{ in.}^2 \; [491 \text{ mm}^2]$$

The maximum tension force based on stress is

$$T = fA = (22)\,(0.785) = 17.27 \text{ kips } [73.7 \text{ kN}]$$

and the maximum force based on elongation is

$$T = \frac{AEe}{L} = \frac{(0.785)\,(29{,}000)\,(1.0)}{(100 \times 12)}$$

$$= 18.97 \text{ kips } [81.8 \text{ kN}]$$

Thus the stress limit is critical.

Net Section and Effective Area

Developing tension in a structural member involves connecting it to something. Some tension-resistive connections—for example, the bolted connection and the threaded connection—reduce the member's load-carrying effectiveness. As a result, designers must determine whether they can afford to sacrifice tension capacity before choosing such a connection.

To insert a bolt, which is a common connection used with members of wood or steel, you must drill or punch a hole in the member. When you create a bolt hole, you get a reduced area, called the *net section*; the unit stress for this area is higher than that at unreduced sections. Of course, the total behavior of a bolted connection is more complex, involving more than the simple tension on the net section, but such a connection often establishes the member's tension capacity.

Another common simple tension member: a round steel rod with spiral threads cut in its ends. To connect two members, insert the rod's ends in holes in the members and screw nuts onto the threaded ends. Cutting the threads reduces the rod's cross section, producing a net section (just as with a bolt hole). However, you can use a rod with forged enlargements at its ends (called *upset ends*); cutting threads in such rods produces a net section whose area is equal to the area of the rod's main part. Upset rods are seldom used for building structures, so the tension rod is typically designed for the reduced net section. *Note:* This consideration for net area also applies to the use of bolts when the load is axial rather than shear.

In steel structures achieving connections sometimes makes it difficult to develop fully the member's tension potential. Figure 9.3 shows a typical connection involving a steel angle: one angle leg is welded to a supporting element. At the connection the tension force is developed only in the angle leg directly grabbed by the welds. Although some tension is developed in the other leg at some distance along the member, designers often ignore the unconnected leg and consider the connected leg to function like a simple bar. The full angle still is effective for stiffness or other considerations, but the tension is limited by the member form and the connection layout.

I discuss the design of connections more in Chapter 11. Keep in mind, however, that achieving the design of tension members depends on how you plan to connect and effectively transfer end tension forces to supporting elements.

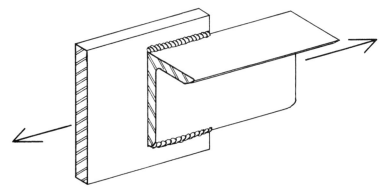

FIGURE 9.3 Typical connection for a steel angle; note that it develops force directly only in the connected leg at the joint.

Flexible Elements

Unlike other structural elements, tension elements are rarely limited by slenderness or aspect ratio. By comparison, columns with height-to-width ratios in excess of 30 or so often are too slender, while beams with span-to-depth ratios over 20 or so usually have critical deflection problems. For tension members, lateral stiffness may literally be zero (e.g., rope or chain) or virtually be zero and negligible (e.g., very long wires, cables, and tie rods).

An excessively slender structural element inevitably develops a form that is a direct response to the only force resistance it can muster—pure tension. If the element has virtually no bending, torsion, or shear capability, you cannot make it develop those internal force actions. The bottom line: A slender element is so severely limited that the designer must determine precisely how loads will develop with such a structure.

On the other hand, the hanger rod, tie rod, or truss member that functions as a simple two-force member simply assumes a straight-line form to resolve the tension force. The only problem is that you must ensure that connection details do not result in something other than a pure axial transfer of tension force to the member. For such members length is not related to stress; in fact, length is limited only (if at all) by considerations for elongation, sag, or vibration in a low-tension state.

Stiff tension elements may function in pure axially stressed situations; but they also have some capacity for other actions and may develop combined stress action.

A superflexible tension element cannot assume the rigid, minutely deflected form of a beam. Instead it assumes a profile that permits it to act essentially in pure tension. However, this profile must be "honest"—that is, not concocted by the designer, but one that the loads and the loaded structure actually can achieve.

Spanning Cables

The steel superflexible tension element known as *cable* consists of bundled steel wires. *Note:* Technically, whether you call such an element a cable, rope, or strand depends on its form. For this discussion I use the familiar name "cable," although the structural element I detail here is actually a strand.

The single-span cable in Figure 9.4a spans horizontally and supports only its own dead weight. The natural draped shape

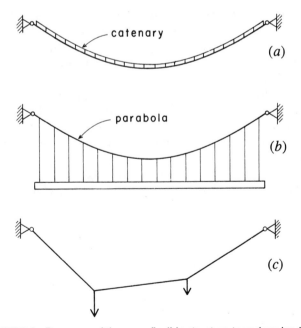

FIGURE 9.4 Response of the superflexible structure to various loadings.

assumed by the cable is a catenary curve, whose profile is described by the equation

$$y = \frac{a}{2} \left(e^{x/a} + e^{-x/a} \right)$$

This form is not particularly useful except for cables that carry only their own weight (such as electrical transmission lines) or whose loads are proportionally small compared to their weights.

In this section's problems you can ignore the weight of the cable without introducing significant error. The cable profile thus is a pure response to the loads' static resolution:

- a simple parabola (Figure 9.4b) when loaded with a uniformly distributed load
- a form consisting of straight segments (Figure 9.4c) when loaded with individual concentrated loads

The cable in Figure 9.5a supports a single concentrated load (W) and has four exterior reaction components (H_1, V_1, H_2, and V_2). Because the cable is indeterminate based on static equilibrium conditions alone, you can analyze the structure only by using the fact that the cable cannot develop bending resistance and therefore no internal bending moment can exist at any point. Also note that individual cable segments operate as two-force members; thus the direction of T_1 must be the same as the slope of the left cable segment and the direction of T_2 must be the same as the slope of the right segment.

Consider the free-body diagram of the whole cable (Figure 9.5b): if you take moments about support 2, assuming clockwise moment as positive, then

$$\Sigma M_2 = -(W \times b) + (V_1 \times L) = 0$$

from which

$$V_1 = \frac{b}{L} W$$

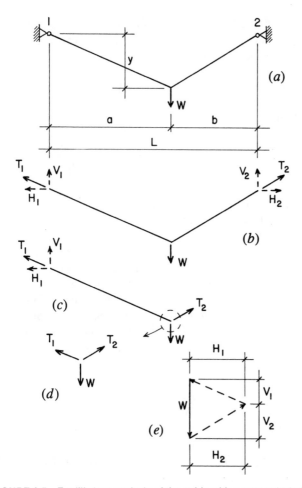

FIGURE 9.5 Equilibrium analysis of the cable with a concentrated load.

Similarly, using moments about support 1,

$$V_2 = \frac{a}{L} W$$

Consider the free-body diagram of the cable's left portion (Figure 9.5c): if you take moments about the point of the load, assuming clockwise moment as positive, then

$$\Sigma M = 0 = +(V_1 \times a) - (H_1 \times y)$$

from which

$$H_1 = \frac{a}{y} V_1$$

Referring again to the free-body diagram for the entire cable (Figure 9.5b), note that the two horizontal reaction components are the only horizontal forces (thus $H_1 = H_2$).

You now know enough to determine the four reaction components and the actual values of T_1 and T_2. Also note that the single load and the two cable tension forces form a simple concentric force system at the point of the load (see Figure 9.5d). You can analyze this system graphically by constructing the force triangle shown in Figure 9.5e. If desired, you can project the values for the reaction components from the force vectors for the two cable forces as shown.

Example 2. Find the horizontal and vertical components of the reactions and the tension forces in the cable for the system shown in Figure 9.6a.

Solution: Using the relationships just derived for the structure in Figure 9.5,

$$V_1 = \frac{b}{L} W = \frac{8}{13} (100) = 61.54 \text{ lb}$$

$$V_2 = \frac{a}{L} W = \frac{5}{13} (100) = 38.46 \text{ lb}$$

$$H_1 = H_2 = \frac{a}{y} V_1 = \frac{5}{6} (61.54) = 51.28 \text{ lb}$$

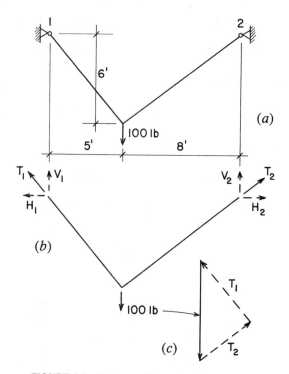

FIGURE 9.6 Reference figure for Example 2.

$$T_1 = \sqrt{(V_1)^2 + (H_1)^2} = \sqrt{(61.54)^2 + (51.28)^2} = 80.1 \text{ lb}$$

$$T_2 = \sqrt{(V_2)^2 + (H_2)^2} = \sqrt{(38.46)^2 + (51.28)^2} = 64.1 \text{ lb}$$

When the two supports are not at the same elevation, the preceding problem is more complex. The solution is still determinate, however, as I show in the following example.

Example 3. Find the horizontal and vertical components of the reactions and the tension forces in the cable for the system shown in Figure 9.7a.

Solution: From the free-body diagram in Figure 9.7b, note that

$$\Sigma F_v = 0 = V_1 + V_2 + 100, \quad \text{thus } V_1 + V_2 = 100$$

$$\Sigma F_h = 0 = H_1 + H_2, \quad \text{thus } H_1 + H_2$$

Consider clockwise moments about point 2 as positive:

$$\Sigma M_2 = 0 = +(V_1 \times 20) - (H_1 \times 2) - (100 \times 12)$$

From the moment equation,

$$20\ V_1 - 2\ H_1 = 1200$$

From the geometry of T_1, observe that

$$H_1 = \frac{8}{6}\ (V_1)$$

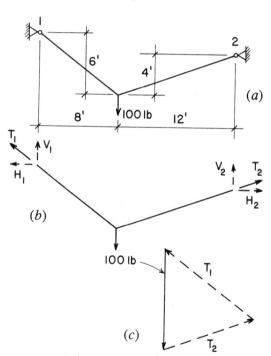

FIGURE 9.7   Reference figure for Example 3.

Substituting in the moment equation,

$$20\,V_1 - (2)\,\frac{8}{6}\,(V_1) = 1200$$

$$\frac{104}{6}\,V_1 = 1200$$

$$V_1 = \frac{6}{104}\,(1200) = 69.23\text{ lb}$$

Then

$$H_1 = \frac{8}{6}\,(V_1) = \frac{8}{6}\,(69.23) = 92.31\text{ lb} = H_2$$

$$V_2 = 100 - V_1 = 100 - 69.23 = 30.77\text{ lb}$$

$$T_1 = \sqrt{(69.23)^2 + (92.31)^2} = 115.4\text{ lb}$$

$$T_2 = \sqrt{(30.77)^2 + (92.31)^2} = 97.3\text{ lb}$$

A cable loaded with a uniformly distributed load along a horizontal span (see Figure 9.8a), assumes a simple parabolic (second-degree) curve profile. From the free-body diagram in Figure 9.8b, observe that the horizontal component of the internal tension is the same for all points along the cable, due to the equilibrium of horizontal forces. The vertical component of the internal tension, meanwhile, varies as the slope of the cable changes, maxing out at the support and decreasing to zero at the center of the span. The maximum internal tension thus occurs at the support and the minimum internal tension occurs at the center of the span.

From Figure 9.8a,

$$V_1 = V_2 = \frac{wL}{2}$$

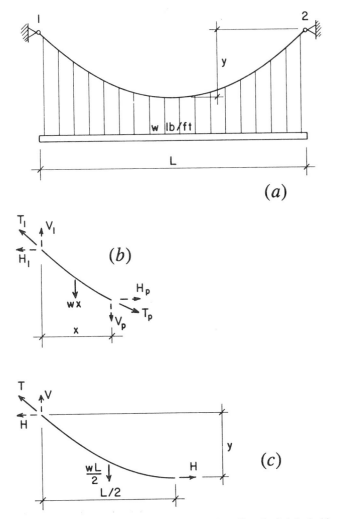

FIGURE 9.8   Behavior of the cable loaded with uniformly distributed load.

From Figure 9.8c, which is a free-body diagram of the cable's left half, summing the moments about the left support yields

$$\Sigma M = 0 = (H_c \times y) + \left(\frac{wL}{2} \times \frac{L}{4} \right)$$

Thus the formula for horizontal force at all points is

$$H_c = \frac{wL^2}{8y}$$

By using the equations for the vertical reaction component and the horizontal force (which is also the horizontal reaction component), you can determine the tension at the supports.

Combined Action: Tension Plus Bending

In some situations, both an axial tension force and a bending moment occur at the same cross section.

Consider the hanger shown in Figure 9.9*a*. A 2-in.-square steel bar is welded to a plate, and the plate is bolted to the bottom of a wood beam. Another steel plate (one with a hole) is welded to the face of the bar, and a load is hung from the hole. In this situation the steel bar faces combined actions of tension and bending, both produced by the hung load (Figure 9.9*b*). The bending moment is the product of the load and its eccentricity from the centroid of the bar cross section:

$$M = 5000 \times 2 = 10{,}000 \text{ in.-lb.}$$

$$[22 \times 50 = 1100 \text{ kN-m}]$$

For this case, find the two stresses separately and then add them.

For the direct tension effect (Figure 9.9*c*),

$$f_a = \frac{N}{A} = \frac{5000}{4} = 1250 \text{ psi } [8.8 \text{ MPa}]$$

For the bending stress, the section modulus of the bar is

$$S = \frac{bd^2}{6} = \frac{2(2)^2}{6} = 1.333 \text{ in.}^3 \ [20.82 \times 10^3 \text{ MPa}]$$

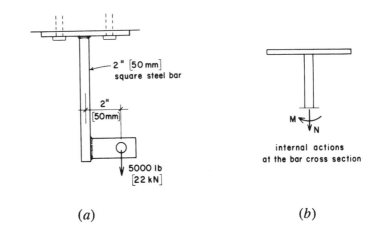

(a)   (b)

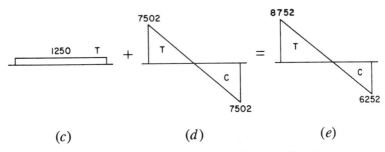

(c)   (d)   (e)

FIGURE 9.9 Development of combined bending and direct tension.

Then, for the bending stress (Figure 9.9*e*),

$$f_b = \frac{M}{S} = \frac{10{,}000}{1.333} = 7502 \text{ psi } [52.8 \text{ MPa}]$$

The stress combinations are (Figure 9.9*e*)

maximum f = 1250 + 7502 = 8752 psi [61.6 MPa] (tension)
minimum f = 1250 − 7502 = 6252 psi [44.0 MPa]
    (compression)

The reversal compression stress is less than the maximum tension stress, but it is critical in some situations. Although the bar in this example probably can develop the compression, other member cross-sections are not so versatile. A thin bar, for example, may become critical in buckling due to the compression, even though the tension stress is higher.

Problem 9.3.A,B. Find the maximum tension in the cable if T = 10 kips (see Figure 9.10) when (a) $x = y = 10$ ft, $s = t = 4$ ft (b) $x = 12$ ft, $y = 16$ ft, $s = 8$ ft, $t = 12$ ft.

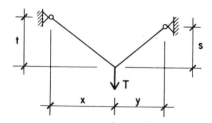

FIGURE 9.10   Reference figure for Problems 9.3.A. and 9.3.B.

10

STEEL TRUSSES

Trussing, or triangulated framing, is a means for developing structural stability. Another advantage: Designers use trusses to produce very light two- and three-dimensional structural elements.

In this chapter I discuss some uses of simple, planar trusses for building structures. I focus on roof trusses—an application that capitalizes most on the truss' light weight and freedom of form.

10.1 GENERAL CONSIDERATIONS

Traditionally designers use the truss to achieve the simple, double-slope, gabled roof form. Such trussing combines sloping members and a horizontal bottom member, as shown in Figure 10.1. The span size determines how designers may fill the simple triangle.

Truss components include the following:

Chord Members. The top and bottom boundary members, akin to a steel beam's top and bottom flanges. For modest-size trusses, such a member often consists of a single element con-

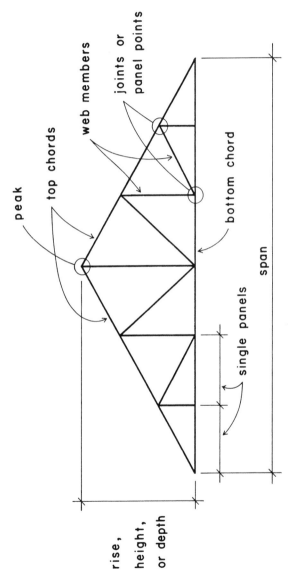

FIGURE 10.1 Truss elements.

tinuous through several joints; its total length is limited only by the maximum piece you can obtain from a supplier.

Web Members. The interior members. Unless you add interior joints, these members consist of a single piece between chord joints.

Panels. A repetitive modular unit. Joints are sometimes known as panel points.

Critical to a truss is its overall height, sometimes referred to as *rise* or *depth*. For the truss in Figure 10.1, this dimension relates to the roof pitch; it also determines the length of the web members. *Note:* Critical to a truss' efficiency as a spanning structure is its ratio of span to height. Although beams and joists may function with span/height ratios as high as 20 to 30, trusses generally require much lower ratios.

Designers use trusses in a number of ways. For example, Figure 10.2 shows a series of single-span, planar trusses together with the

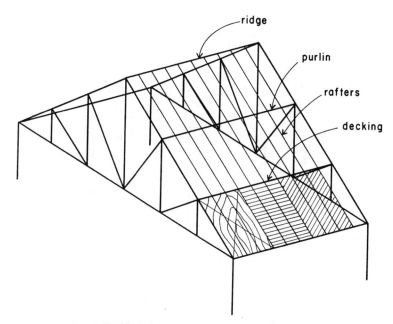

FIGURE 10.2 A roof structure with trusses.

structural elements that develop the roof system. Figure 10.3, meanwhile, shows a series of trusses with parallel chords; designers often use this system for a floor or a flat roof.

When trusses are widely spaced, designers often use purlins to span between the trusses. The purlins are usually supported at the trusses' top chord joints to avoid bending in the chords; the purlins, in turn, support a series of closely spaced rafters parallel to the trusses. Because the roof deck is attached to the rafters, the roof surface actually floats above the trusses. When the trusses are closely spaced, you may eliminate the purlins and increase the size of the top chords to accommodate the additional bending caused by the rafters. If the trusses are jammed together, you may eliminate the rafters as well and place the deck directly on the trusses' top chords.

In some situations the system requires additional elements. For example, if the structure needs a ceiling, add another framing system at or below the bottom chords. You also may need to use some bracing system perpendicular to the trusses.

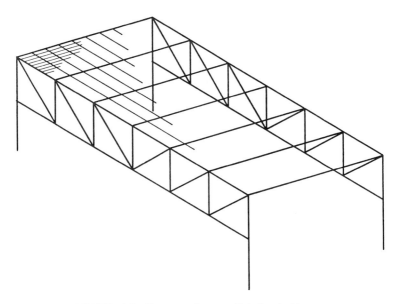

FIGURE 10.3 Flat-spanning, parallel-chorded trusses.

Figure 10.4 pictures ten common truss forms. (The names of classic truss patterns are standard structural terms.) The two most common forms of small and medium-size steel trusses are shown in Figure 11.17. In both cases you may connect the members by bolts, or welds. Typically, designers use welds for shop connections, high-strength bolts (torque tensioned) for permanent field connections, and unfinished bolts for temporary field connections.

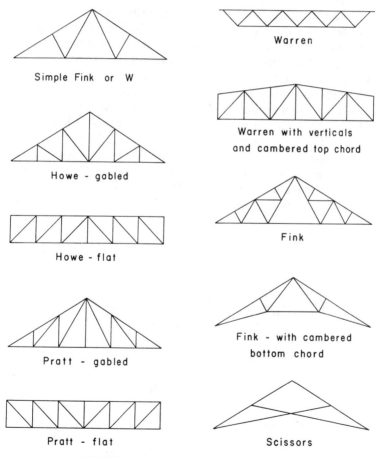

FIGURE 10.4 Common steel truss patterns.

10.2 BRACING FOR TRUSSES

Planar trusses are very thin structures and require some form of lateral bracing. In turn, you must design the compression chord for its laterally unbraced length. If there is no lateral bracing, the chord's unbraced length in a direction perpendicular to the plane of the truss is the full truss length. Obviously, you cannot design a slender compression member for this unbraced length. *Note:* The chord is braced by other truss members at each joint in the plane of the truss.

In most buildings other construction elements provide some or all the necessary bracing for the trusses. In Figure 10.5a, the truss' top chord is braced at each truss joint by the purlins. If the roof deck is a reasonably rigid planar structural element and is adequately attached to the purlins, this bracing is adequate for the compression chord. However, you also must brace the truss generally for out-of-plane movement throughout its height. In Figure 10.5a such bracing is provided by a vertical plane of X-bracing at every other truss panel point (note that the purlin also helps out). Because one panel of this bracing can brace a pair of trusses, you can theoretically place such bracing only in alternate bays. However, if the bracing is part of the building's general bracing system it may need to be continuous.

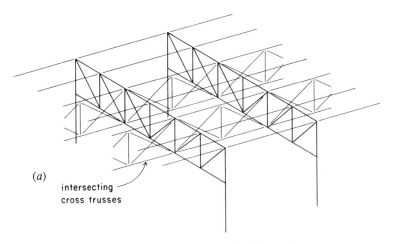

(a)

intersecting
cross trusses

FIGURE 10.5 Forms of lateral bracing for steel trusses.

Usually light trusses that directly support a deck (see Figure 10.5b) are braced adequately by the deck. Because such bracing is continuous, the chord's unbraced length may be virtually zero (depending on how the deck is attached). In this situation you often can limit additional bracing to a series of continuous steel rods or single small angles that are attached to the bottom chords.

In Figure 10.5c a horizontal plane of X-bracing sits between two trusses at the level of the bottom chords. You can use this single braced bay to brace several other bays of trusses by connecting them to the X-braced trusses with horizontal struts. Meanwhile, the top chord is braced by the roof construction. Such bracing often is part of the building's general lateral bracing system.

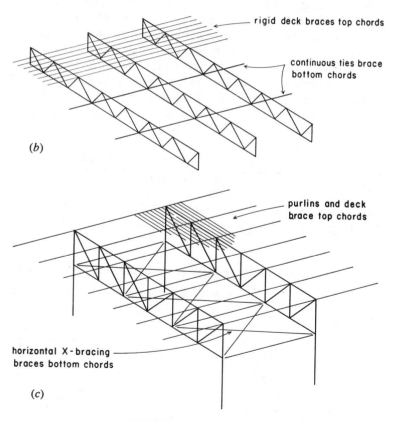

rigid deck braces top chords

continuous ties brace bottom chords

(b)

purlins and deck brace top chords

horizontal X-bracing braces bottom chords

(c)

FIGURE 10.5 *Continued*

10.3 LOADS ON TRUSSES

When designing a roof truss, you first must compute the dead and live loads the truss must support. Dead load is weight of all construction materials supported by the truss—roof covering and roof deck, purlins and sway bracing, ceiling and any suspended loads, and the truss itself (Table 12.1 gives the weights of certain roofing materials). Live load includes loads resulting from snow, wind, occupancy (i.e., loads experienced during roof construction and maintenance), and ponding.

Dead loads are downward vertical forces, so the end reactions of the truss are also vertical. Table 10.1 provides estimated weights of steel trusses for various spans and pitches (only after you design the truss can you compute its actual weight). Should you need a weight not listed, estimate a load in pounds per square foot of roof surface and then consider this load as acting at the panel points of the upper chord. A more exact method is to apportion a part of such loads to the panel points of the lower chord, but such precision is customary only for trusses with exceptionally long spans.

Required live loads are specified by local building codes. Snow load depends primarily on anticipated snow accumulation and roof slope. Freshly fallen snow may weigh as much as 10 lb per ft^3 [0.13 kg/m^3] and accumulations of wet or packed snow may weigh even more. How much snow stays on a roof over a given period depends on the roof type as well as the slope—for example, snow slides off a metal or slate roof more readily than off a wood shingle surface. (The amount of insulation in the roof construction also influences the period of retention as escaping heat melts the snow.)

TABLE 10.1 Approximate Weight of Steel Trusses in Pounds per Square Foot of Supported Roof Surface

Span		Slope of Roof			
ft	m	45°	30°	20°	Flat
Up to 40	Up to 12	5	6	7	8
40–50	12–15	6	7	7	8
50–65	15–20	7	8	9	10
65–80	20–25	8	9	10	11

The basic live load, based on the total roof surface area supported by the structure, can be modified when the roof is significantly sloped. Table 12.2 gives the minimum roof live loads specified by the *Uniform Building Code*, 1994 ed. (Ref. 1); however, note that these loads apply when snow load is not the critical concern.

Note: Local building codes specify wind design requirements. For a general explanation of how to analyze and design for wind and earthquake effects, refer to *Simplified Building Design for Wind and Earthquake Forces* (Ref. 9).

10.4 INVESTIGATION FOR INTERNAL FORCES IN PLANAR TRUSSES

In this section I illustrate how to determine the internal forces in the truss—that is, the tension and compression forces in individual truss members.

Graphical Analysis

Figure 10.6 shows a single-span, planar truss subjected to vertical gravity loads.

The space diagram shows the truss form, the support conditions, and the loads. The letters between loads and individual truss members enable you to identify individual forces at the truss joints: each force at a joint is denoted by a two-letter combination.

The separated joint diagram illustrates not only the complete force system at each joint, but also how the joints interrelate. Each force at each joint is designated by a two-letter symbol that is obtained by reading clockwise around the joint in the space diagram. Note that the two-letter symbols are reversed at the opposite ends of each truss member; for example, the top chord member at the left end of the truss is designated *BI* when shown in the joint at the left support (joint 1) and *IB* when shown in the first interior upper chord (joint 2).

The Maxwell diagram is a composite force polygon for the truss' external and internal forces. Named after one of its original users, the English engineer James Clerk Maxwell, this diagram con-

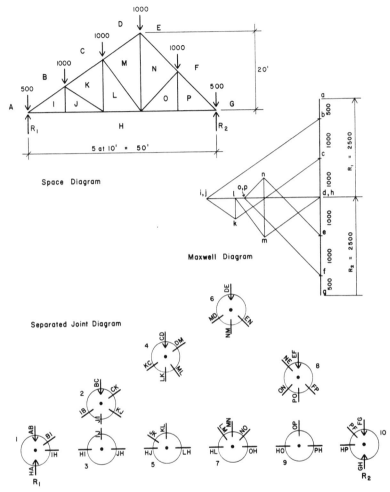

FIGURE 10.6 Diagrams used to analyze a planar truss.

stitutes a complete solution for the magnitudes and senses of the truss' internal forces. You construct this diagram as follows:

1. Construct the Force Polygon for the External Forces

First find the values for the reactions, using either graphical or algebraic techniques. (Algebraic techniques are usually much simpler and faster.) In this example, the loading is symmetrical: each reaction is equal to one-half the total load on the truss, or 5000/2 = 2500 lb.

The external forces in this case are all in a single direction, so the force polygon for the external forces is a straight line. Given the two-letter symbols for the forces, read the force sequence clockwise around the truss. The loads are thus *AB, BC, CD, DE, EF,* and *FG;* the two reactions are *GH* and *HA.* The external force vector sequence is read as *A* to *B, B* to *C, C* to *D,* and so on—returning to *A* at the end shows that the force polygon closes and that the external forces are in static equilibrium. (I pulled the vectors for the reactions to the side to indicate them more clearly.)

Note that the space diagram uses uppercase letters, whereas the Maxwell diagram uses lowercase letters. An alphabetic correlation is retained (*A* to *a*), while any possible confusion between the two diagrams is prevented. The letters on the space diagram designate spaces, while the letters on the Maxwell diagram designate points of intersection.

2. Construct the Force Polygons for the Individual Joints

After you identify the external force vectors, you must locate on the Maxwell diagram points that correspond to any remaining space diagram letters. To locate these points, use two relationships. The first: Truss members can resist only forces that are parallel to the members' positioned directions. Thus, from the space diagram, you know the directions (angles) of all the internal forces. The second: The intersection of two lines is a point. For example, consider the forces at joint 1, as shown in the separated joint diagram. Note that two of the four forces are known (the load and the reaction) and two are unknown (the internal forces in the truss members). The force polygon for this joint, is read as *ABIHA.* In this polygon, *AB* represents the load, *BI* the force in the upper chord member, *IH* the force in the lower chord member, and *HA* the reaction. Thus you can determine the location of point *I* by noting that *I* must be in a horizontal direction from *H* (corresponding to the horizontal position of the lower chord) and in a direction from *B* that is parallel to the position of the upper chord. In sum, you can use two known points to project lines of known direction; the intersection of those lines determines the location of another point.

To find the remaining points on the Maxwell diagram, use the same process. Once you find all the points, the diagram is complete and you can use it to find the magnitude and sense of each internal force.

To construct a Maxwell diagram, simply move from joint to joint. The only limitation: It is impossible to find more than one unknown point for any single joint. For example, if you try to solve joint 7 on the separated joint diagram in Figure 10.6, knowing only the locations of letters *A* through *H*, you must locate four unknown points (*L*,*M*,*N* and *O*), or three more unknowns than you can determine in a single step. As a result, you must first solve for three of the unknowns by using other joints.

Note: Solving for an unknown point corresponds to finding two unknown forces at a joint, because each letter on the space diagram is used in two internal force identifiers. Two unknowns are the maximum that can be solved for in the equilibrium of a coplanar, concurrent force system.

From a completed Maxwell diagram, you can investigate the following for any internal force:

- Determine the magnitude by measuring the length of the line using the scale that was used to plot the vectors for the external forces.
- Determine the sense by comparing the forces around a single joint in the space diagram to the same letter sequences on the Maxwell diagram.

Figure 10.7*a* shows the force system at a joint and the corresponding force polygon—the known forces are represented by solid lines on the force polygon, the unknown forces by dashed lines. Reading clockwise from the force system, the forces are *AB*, *BI*,*IH*, and *HA*. To read from *a* to *b* on the force polygon is to move in the order of the force sense (i.e., from tail to head of the force vector that represents the external load on the joint). If you continue in this sequence, this force sense flow is continuous. Thus to read from *b* to *i* is to read from tail to head of the force vector, indicating that force *BI* has its head at the left end. When you transfer this sense indication to the joint diagram, you discover that force *BI* is in compression—that is, it pushes, rather than pulls, on the joint.

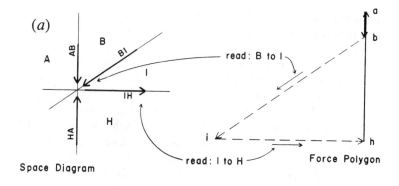

(a)

Space Diagram

Force Polygon

read: B to I

read: I to H

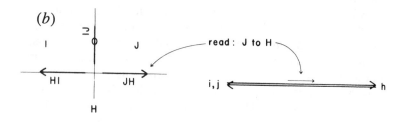

(b)

read: J to H

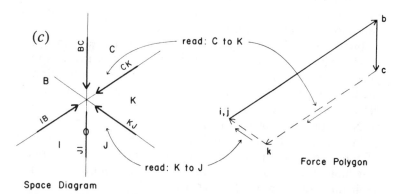

(c)

Space Diagram

Force Polygon

read: C to K

read: K to J

FIGURE 10.7 Graphical solutions for individual truss joints (refer to Fig. 10.6 for joint numbering): (a) Joint 1 (b) Joint 3 (c) Joint 2.

Reading from *i* to *h* on the force polygon, meanwhile, you can see that the arrowhead for this vector is on the right, which translates to a tension effect.

The more forces you know, the easier it is to solve for adjacent joints. However, be careful to note that senses reverse at the opposite ends of the members. For example, from the separated joint diagram in Figure 10.6, if the upper chord member shown as force *BI* (joint 1) is in compression, its vector points down and to the left (see Figure 10.7a). However, when the same force is shown as *IB* (joint 2), its effect is reversed, so its vector points up and to the right. Similarly, the tension effect of the lower chord is indicated in joint 1 by an arrowhead on the right end of the force *IH*, but the same tension force in joint 3 is indicated by an arrowhead on the left end of the force *HI*.

After you solve for joint 1, you can transfer the known force in the upper chord to joint 2. Thus to solve for joint 2, you need find only three unknowns—the load *BC* and the chord force *IB* are now known. Of course, you still cannot solve joint 2 because two unknown points (*k* and *j*) correspond to three unknown forces. You can, however, proceed to joint 3, at which only two forces are unknown. You may find the point *j* by projecting vector *IJ* vertically from *i* and vector *JH* horizontally from *h*. In fact, because *i* is located horizontally from *h*, you know that *j* also must be located horizontally from *h*. In turn, you know that the vector *IJ* has zero magnitude, no stress exists in this truss member for this loading condition, and points *i* and *j* are coincident.

Figure 10.7b shows the joint force diagram and the force polygon for joint 3. In the joint force diagram, a zero, not an arrowhead, goes on the vector line for *IJ* to indicate the zero stress condition. In the force polygon, the two force vectors are slightly separated for clarity, but they are actually coincident.

Now you can proceed to joint 2, since only two forces at this point remain unknown. From Figure 10.7c read the force polygon after reading clockwise around the joint. Following the continuous direction of the force arrows in the sequence *BCKJIB*, you can establish the sense for the forces *CK* and *KJ*.

The bottom line: You can start at one end and work across the truss from joint to joint to construct a Maxwell diagram. In this example, you can locate points on the Maxwell diagram in the sequence *i-j-k-l-m-n-o-p*, solving the joints in the sequence 1,3,2,5,4,6,7,9,8. However, you can minimize error by working from both ends of the truss. Thus a better procedure would be to find points *i-j-k-l-m*, working from the left end of the truss, and then to find points *p-o-n-m*, working from the right end. *Note:* The difference in the two locations for *m* constitutes the error in drafting accuracy.

Problem 10.4.A,B. Using a Maxwell diagram, find the internal forces in the truss pictured in Figure 10.8.

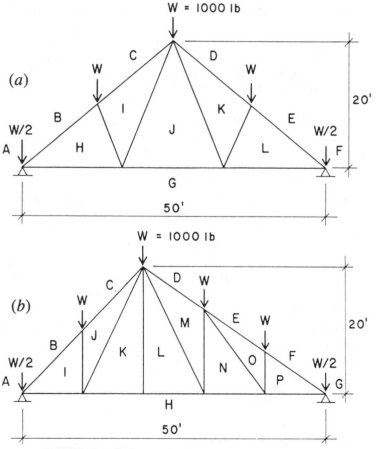

FIGURE 10.8 Reference for Problems 10.4.A. and 10.4.B.

Algebraic Analysis

You also can use algebra to solve the internal forces in a truss. In the *method of joints*, you solve the concentric force systems at the individual joints by using simple force equilibrium equations and known geometric conditions. *Note:* I demonstrate this method using the same truss illustrated in Figure 10.6. First determine the loads and the forces at the supports (the reactions). Then determine the internal forces. When considering the equilibrium of the individual joints, note that the concentric force systems permit you to solve for only two unknowns at a time at any joint. If more unknowns exist at a joint, you must solve other joints first.

I urge you to draw the force system at the joint, including the sense (algebraic sign or arrowed placement) and magnitude of known forces; then add the unknown forces, represented by lines without arrowheads (unless it is clear what their sense is). To simplify the mathematical process, replace forces in members that are not vertical or horizontal with their vertical and horizontal components. Then consider the equilibrium of the joint by using the two conditions for the joint: the sum of the vertical forces equals zero and the sum of the horizontal forces equals zero.

The method of joints determines force sense automatically. However, I recommend that, whenever possible, you predict the sense of unknown forces by observing the joint conditions, as I illustrate in the following solution for the force *BI*. (The problem I solve is shown in Figure 10.9a.)

First break down the force into its vertical and horizontal components (see Figure 10.9b). Although this step increases the actual number of unknowns, it makes an equilibrium summation easier.

The condition for vertical equilibrium is shown in Figure 10.9c. Obviously the vertical component must act downward; that is, *BI* pushes on the joint with a compression force. Voilà, I just identified the force sense. An algebraic equation for the vertical force summation (with upward force considered positive) is

$$\Sigma F_v = 0 = +2500 \ -500 \ -BI_v$$

From this equation, BI_v is 2000 lb.

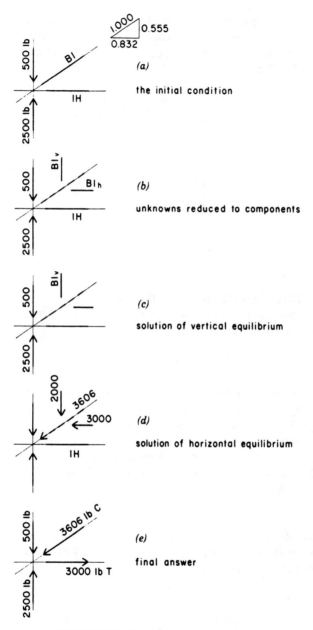

FIGURE 10.9   Algebraic solution for Joint 1.

From the geometry of BI (see Figure 10.9a), you can establish a relationship between the force and its components. In other words, you know if any one force or component, you can determine the others. Thus

$$\frac{BI}{1.000} = \frac{BI_v}{0.555} = \frac{BI_h}{0.832}$$

Then

$$BI_h = \frac{0.832}{0.555}\,(2000)\ =\ 3000\ \text{lb}$$

$$BI = \frac{1.000}{0.555}\,(2000)\ =\ 3606\ \text{lb}$$

Figure 10.9d shows the results of this analysis.

The condition for horizontal equilibrium (with force sense to the right considered positive) is

$$\Sigma\,F_h\ =\ 0\ =\ IH\ -\ 3000$$

From this equation you can find that the force in IH is 3000 lb and acts away from the joint as a tension force.

Note: The complete system is shown in Figure 10.9e.

You now can solve for the forces at Joint 3. As you can see from Figure 10.10a, it is impossible to have a force in member IJ because no other force exists to oppose it. Stated algebraically,

$$\Sigma\,F_v\ =\ 0\ =\ IJ$$

Note also that the force in member JH must be equal in value and opposite in sense from the force in the only other horizontal member at the joint. Stated algebraically,

$$\Sigma\,F_h\ =\ 0\ =\ JH\ -\ 3000$$

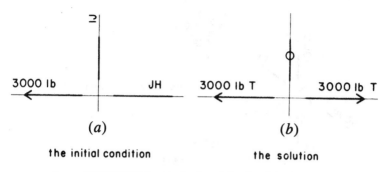

(a)

the initial condition

(b)

the solution

FIGURE 10.10 Algebraic solution for Joint 3.

You now can solve Joint 2 (see Figure 10.11a). Of the five forces at the joint, only two remain unknown in sense and value, so the joint is statically determinate. Because these two unknown forces are in sloping members, first resolve them into vertical and horizontal components, as in Figure 10.11b. To simplify the algebraic equations for this joint, assume a sense for each force; if the solution is positive, the assumption is correct, whereas a negative value indicates a true sense opposite of the assumption. For this demonstration, I assume the force senses shown in Figure 10.11c.

Considering the equilibrium of vertical forces (see Figure 10.11d), the summation is

$$\Sigma F_v = 0 = -1000 + 2000 - CK_v - KJ_v$$

By using the geometry of the forces,

$$1000 - 0.555\, CK - 0.555\, KJ = 0$$

Similarly, a summation for horizontal forces (see Figure 10.11e) yields

$$\Sigma F_h = 0 = +3000 - CK_h + KJ_h$$
$$3000 - 0.832\, CK + 0.832\, KJ = 0$$

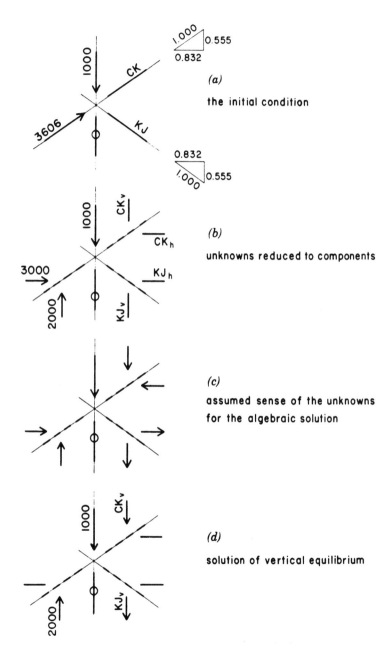

FIGURE 10.11 Algebraic solution for Joint 2.

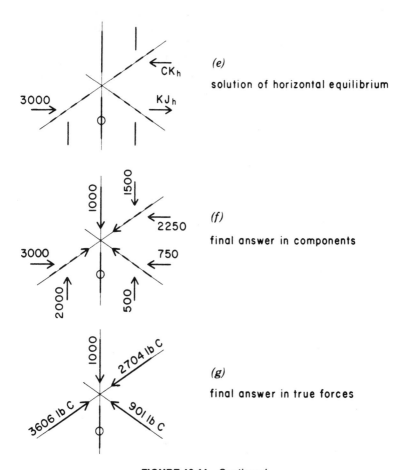

FIGURE 10.11 *Continued*

Simultaneously solving these equations yields values of $+2704$ lb for CK and -901 lb for KJ. The plus value for CK indicates that the assumed sense (Figure 10.11c) was correct, while the negative value for KJ was incorrect. If so desired, you may determine the values for these components from their geometry.

Figure 10.11f is a resolution of the joint for all vertical and horizontal forces, and Figure 10.11g is a resolution of the true forces.

After you know all the internal forces, you may record or display the results in a number of ways. The most direct means is to display the force values and senses on a scaled diagram of the truss, as shown in the upper part of Figure 10.12, where *T* stands for tension and *C* for compression. A zero stress member is indicated by a zero on the member.

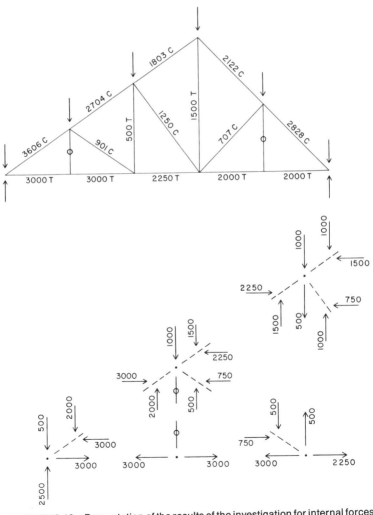

FIGURE 10.12   Presentation of the results of the investigation for internal forces in the truss.

If you used the method of joints, you may record the results on a separated joint diagram, as shown in the lower part of Figure 10.12.

Problem 10.4.C,D. Using the method of joints, find the internal forces in the truss pictured in Figure 10.8.

Internal Forces Found By Coefficients

Figure 10.13 shows ten simple trusses. Table 10.2 lists coefficients that you may use to solve for the internal forces in these trusses. For the gabled trusses, coefficients are given for three slopes of the top chord: 4 in 12, 6 in 12, and 8 in 12. For the parallel-chorded trusses,

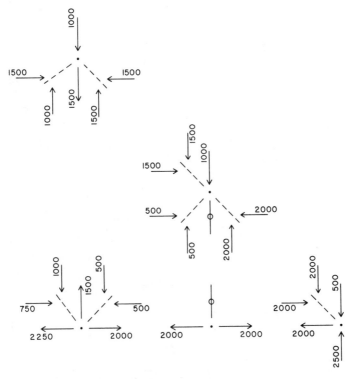

FIGURE 10.12 *Continued*

TABLE 10.2 Coefficients for Internal Forces in Simple Trusses[a]

Force in members = (table coefficient) X (panel load, W)

T indicates tension, C indicates compression

Gable Form Trusses

Truss Member	Type of Force	Roof Slope 4/12	6/12	8/12
Truss 1 – Simple Fink				
AD	C	4.74	3.35	2.70
BE	C	3.95	2.80	2.26
DC	T	4.50	3.00	2.25
FC	T	3.00	2.00	1.50
DE	C	1.06	0.90	0.84
EF	T	1.06	0.90	0.84
Truss 2 – Fink				
BG	C	11.08	7.83	6.31
CH	C	10.76	7.38	5.76
DK	C	10.44	6.93	5.20
EL	C	10.12	6.48	4.65
FG	T	10.50	7.00	5.25
FI	T	9.00	6.00	4.50
FM	T	6.00	4.00	3.00
GH	C	0.95	0.89	0.83
HI	T	1.50	1.00	0.75
IJ	C	1.90	1.79	1.66
JK	T	1.50	1.00	0.75
KL	C	0.95	0.89	0.83
JM	T	3.00	2.00	1.50
LM	T	4.50	3.00	2.25
Truss 3 – Howe				
BF	C	7.90	5.59	4.51
CH	C	6.32	4.50	3.61
DJ	C	4.75	3.35	2.70
EF	T	7.50	5.00	3.75
EI	T	6.00	4.00	3.00
GH	C	1.58	1.12	0.90
HI	T	0.50	0.50	0.50
IJ	C	1.81	1.41	1.25
JK	T	2.00	2.00	2.00
Truss 4 – Pratt				
BF	C	7.90	5.59	4.51
CG	C	7.90	5.59	4.51
DI	C	6.32	4.50	3.61
EF	T	7.50	5.00	3.75
EH	T	6.00	4.00	3.00
EJ	T	4.50	3.00	2.25
FG	C	1.00	1.00	1.00
GH	T	1.81	1.41	1.25
HI	C	1.50	1.50	1.50
IJ	T	2.12	1.80	1.68

Flat-Chorded Trusses

Truss Member	Type of Force	6 Panel Truss $\frac{h}{p}=1$	$\frac{h}{p}=\frac{3}{4}$	8 Panel Truss $\frac{h}{p}=1$	$\frac{h}{p}=\frac{3}{4}$
Truss 5 – Pratt					
BI	C	2.50	3.33	3.50	4.67
CK	C	4.00	5.33	6.00	8.00
DM	C	4.50	6.00	7.50	10.00
EO	C	–	–	8.00	10.67
GH	O	0	0	0	0
GJ	T	2.50	3.33	3.50	4.67
GL	T	4.00	5.33	6.00	8.00
GN	T	–	–	7.50	10.00
AH	C	3.00	3.00	4.00	4.00
IJ	C	2.50	2.50	3.50	3.50
KL	C	1.50	1.50	2.50	2.50
MN	C	1.00	1.00	1.50	1.50
OP	C	–	–	1.00	1.00
HI	T	3.53	4.17	4.95	5.83
JK	T	2.12	2.50	3.54	4.17
LM	T	0.71	0.83	2.12	2.50
NO	T	–	–	0.71	0.83
Truss 6 – Howe					
BH	O	0	0	0	0
CJ	C	2.50	3.33	3.50	4.67
DL	C	4.00	5.33	6.00	8.00
EN	C	–	–	7.50	10.00
GI	T	2.50	3.33	3.50	4.67
GK	T	4.00	5.33	6.00	8.00
GM	T	4.50	6.00	7.50	10.00
GO	T	–	–	8.00	10.67
AH	C	0.50	0.50	0.50	0.50
IJ	T	1.50	1.50	2.50	2.50
KL	T	0.50	0.50	1.50	1.50
MN	T	0	0	0.50	0.50
OP	O	–	–	0	0
HI	C	3.53	4.17	4.95	5.83
JK	C	2.12	2.50	3.54	4.17
LM	C	0.71	0.83	2.12	2.50
NO	C	–	–	0.71	0.83
Truss 7 – Warren					
BI	C	2.50	3.33	3.50	4.67
DM	C	4.50	6.00	7.50	10.00
GH	O	0	0	0	0
GK	T	4.00	5.33	6.00	8.00
GO	T	–	–	8.00	10.67
AH	C	3.00	3.00	4.00	4.00
IJ	C	1.00	1.00	1.00	1.00
KL	O	0	0	0	0
MN	C	1.00	1.00	1.00	1.00
OP	O	–	–	0	0
HI	T	3.53	4.17	4.95	5.83
JK	C	2.12	2.50	3.54	4.17
LM	T	0.71	0.83	2.12	2.50
NO	C	–	–	0.71	0.83

[a] See Figure 10.13 for truss forms and member identification.

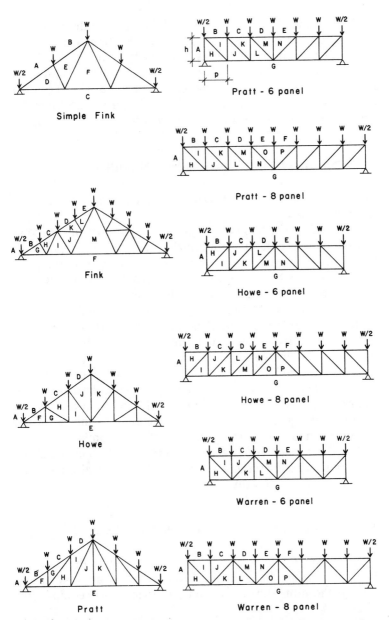

FIGURE 10.13 Simple planar trusses: gabled and parallel-chorded. (*Note:* Reference figure for Table 10.2.)

coefficients are given for two ratios of the truss depth to the truss panel length: 1 to 1 and 3 to 4. Assume that loading, which results from gravity loads, is applied symmetrically to the truss. Table 10.2 values are based on a unit load of 1.0 for W; thus true forces may be simply proportioned for specific W values once actual loading is determined. Because all the trusses are symmetrical, values are given for only half of each truss.

10.5 DESIGN FORCES FOR TRUSS MEMBERS

In truss analysis you must determine for each truss member which forces are critical.

First decide which loading combinations are critical—in some cases numerous potential combinations exist. For example, when both wind and seismic actions are potentially critical and more than one type of live loading occurs (e.g., roof loads plus hanging loads), the number of theoretically possible loadings combinations is overwhelming. However, designers usually can eliminate improbable combinations; for example, it is statistically unlikely that a violent windstorm and a major earthquake shock will occur simultaneously.

After you establish the required design loading conditions, perform separate analyses for each loading. You then can combine the values for each member to ascertain which combination is critical. In some cases you may need to design certain members for one combination and other members, for different combinations.

Note: In most cases design codes permit you to increase allowable stress when designing members whose critical loading includes forces due to wind or seismic loads.

10.6 COMBINED ACTIONS IN TRUSS MEMBERS

When analyzing trusses, designers usually assume that the truss joints handle all the loads. If so, the members are loaded only through the joints and thus face only direct tension or compression forces.

In some cases, however, truss members are directly loaded—for example, when the top chord of a truss supports a roof deck without benefit of joists; the chord member has a linear uniform load and functions as a beam between its end joints.

In such situations you must analyze the whole truss to derive the typical joint loading arrangement. Then design the truss members that sustain the direct loading for the combined effects of the axial force caused by the truss action and the bending caused by the direct loading.

For a typical roof truss, the actual loading consists of the roof load distributed continuously along the top chords and a ceiling loading distributed continuously along the bottom chords—in other words, a combination of axial tension and bending. As a result, you need larger members for both chords, so any estimate of the truss weight must reflect this anticipated additional requirement.

10.7 DESIGN CONSIDERATIONS FOR STEEL TRUSSES

Before designing a steel truss, you must decide whether to use a truss rather than other possible structures. Some conditions that favor using a truss are the following:

- The structure needs an efficient spanning system; for example, in lieu of solid web beams.
- The structure needs a long span, but the load is light (for example, most roofs).
- The structure needs an adaptable roof profile geometry. (In fact, the truss is the most geometrically adaptable structure.)
- The structure needs to be open (for example, to permit air or light to move through the structure, or to allow items such as ducts, piping, wiring, stairs, and so on to pass through the structure).

When designing a truss, you must pay special attention to the following:

- Truss profile and member layout
- Member form
- Connections form (I discuss later in this section)
- Necessary bracing

Note: I discuss all these issues in this chapter. Connections are further discussed in Chapter 11. General truss form is discussed for several cases in Chapter 12. Establishing the connections form is especially critical, relative to construction cost. Of course, the connection form also affects a truss load capacity. Actually, connections relate to almost all truss design issues, including

- *Structural Size.* What works for small trusses may not work for large trusses, which take on greater loads.
- *Truss Member Form.* Different connecting methods are necessary for angles, W shapes, round pipes, tubes, and so on.
- *Truss Layout.* If a design calls for many joints, an expensive connection method may be unfeasible. If many members meet at a single joint, the connection must accommodate the traffic.
- *Supported Elements.* For example, roof, floor, or ceiling structures usually are attached at truss joints; such design complicates a joint's functions.

Because of the wide-ranging truss design concerns, I urge you to see the design problem in as broad a context as possible. To wit, I provide several examples of truss use for the building case studies in Chapter 12. These include:

- Roof trusses for Buildings One and Four
- Open-web steel joist construction for Buildings One, Two, and Three
- Two-way spanning truss system for Building Six

10.8 TWO-WAY TRUSSES

You can turn an ordinary truss system consisting of parallel, planar trusses into a two-way spanning system by connecting the parallel trusses with cross-trussing. However, this system lacks stability in the horizontal plane: the vertical, planar trusses form rectangles, and you need triangulation for stability in the horizontal plane. (Of course, you can rectify this situation by adding horizontal trussing.)

Whereas the planar triangle is the basic unit of the planar truss, the tetrahedron is the basic unit of the spatial truss (see Figure 10.14). A three-dimensional system involves three orthogonal planes (x, y, z coordinate system); the system defines rectangles in each plane and cubes in three dimensions. Triangulating each orthogonal plane produces tetrahedra.

When you use the mutually perpendicular, vertical, planar truss system, you can form squared corners at the system's edge and in relation to plan layouts beneath it. The roof edge is formed by the basic truss system, and the exterior walls meet the truss' bottom at natural truss chord locations.

A purer spatial truss form is based on the tetrahedron's basic spatial triangulation. If all the edges of a tetrahedron are equal in length, the solid form described is not orthogonal—in other words, it does not describe the usual x, y, z system of mutually perpendicular planes (see Figure 10.15). If you use tetrahedrons for a two-way spanning truss system, you may develop a flat structure in the horizontal plane, but the trusses do not inherently develop rectangular plans, but rather triangles, diamonds, and hexagons.

The fully triangulated system can save you money because you can form it with members of the same length and you can use simple joints throughout the truss system. The pure triangulated space system, however, allows you to form truly three-dimensional structures and not just flat, two-spanning systems.

A compromise form—something between the orthogonal and the pure triangular systems—is the *offset grid*, which consists of two horizontal planes constituted as grids of rectangules. In such a sys-

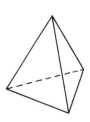

single tetrahedron

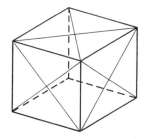

the trussed box

FIGURE 10.14   Three-dimensional trussing: the tetrahedron and the cube.

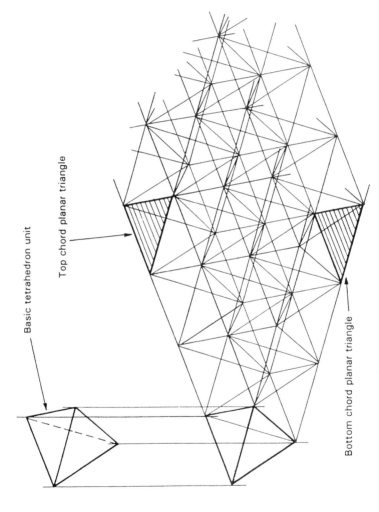

Basic tetrahedron unit

Top chord planar triangle

Bottom chord planar triangle

FIGURE 10.15 A two-way spanning truss system. Note how the tetrahedron is the basic geometric unit. Note also the triangulation produced in the top and bottom planes.

tem, grid intersections in the upper plane (top chord plan) are located over the grid centers in the lower plane (bottom chord plan). Truss web members, which connect the grids, describe vertical planes in diagonal plan directions but do not form vertical planes with the chords. At a typical joint four chord members meet four web diagonal members. The offset grid permits you to develop square plan layouts (usually preferred by designers), while retaining some advantages of the triangulated system: (same-length members, same-form joints).

In sum, although you can use spatial trussing to produce just about any structural unit—including columns, trussed mullions, freestanding towers, beams, surfaces of large multiplaned structures, and even arched surfaces formed as vaults (i.e., cylinders) or domes—truss geometry must relate to structural purpose. By the way, most spatial trusses are steel.

Span and Support Considerations

When planning structures with spatial trussing, you must pay attention to the supports if you want to use the two-way action optimally. The locations of supports define not only the span size, but many aspects of the two-way truss system's behavior.

Figure 10.16 shows possible support systems for a single square panel of a two-way spanning system. In Figure 10.16a support is provided by four columns placed at the outside corners, causing a maximum span condition for the system's interior. This form of support also leads to a very high shear condition in the corner's single quadrant, requiring the edges to act as one-way spanning supports for the two-way system. As a result, the edge chords must be very heavy and the corner web members must accept heavy loads before transferring the vertical force to the columns.

Figures 10.16b and c show support systems that eliminate the edge spanning and corner shear via bearing walls or closely spaced perimeter columns. The trade-off: A lower truss cost, but higher costs for the support system and its foundations. Such support is more restrictive architecturally.

In Figure 10.16d support is provided by four columns placed in the centers of the sides. Such a system requires a considerable edge structure to achieve the corner cantilevers, but it actually reduces

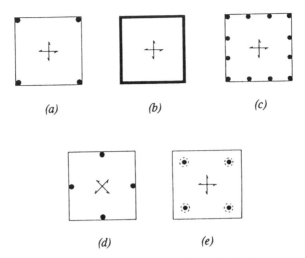

FIGURE 10.16 Various forms of support for a single, two-way-spanning structure.

the span of the interior system. In addition, the overhang effect of the cantilever corners reduces the system's maximum moment. The interior's clear span and the high shear experienced at the four columns are the same as in Figure 10.16*a*.

In Figure 10.16 the columns sit inside the edges —an ideal support system for column shear. This system provides a wider edge spanning strip, divides shear between more truss members, and reduces the interior's clear span to something less than the full truss width. If the exterior wall is at the roof edge, the interior columns may be intrusive. However, if you want a roof overhang, the wall may sit at the column lines; in turn, the structure relates well to the architectural plan.

Two-way spanning units should describe a square. If the support arrangement creates oblong units, the structural action may wind up essentially one-way in function.

Figure 10.17 shows three forms for a span unit—ratio of sides 1:1, 1:1.5, and 1:2. The square unit shares the spanning effort equally in each direction (as long as the system is otherwise fully symmetrical). When the ratio of sides is 1:1.5, the shorter span becomes much stiffer for deflection and attracts as much as 75 percent of the total load. (This ratio is usually the maximum allowable

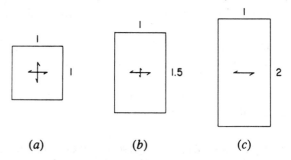

(a)   *(b)*   *(c)*

FIGURE 10.17   Different ratios of spans in a two-way spanning system. As the ratio approaches 2:1, the structure essentially becomes a one-way spanning structure, in spite of support conditions.

to ensure two-way action.) When the ratio of sides is 1:2, the only load carried in the long direction is that near the ends of the plan unit, adjacent to the short sides. The narrow unit bends mostly in single, archlike form, a far cry from the two-way curvature (domed or dishlike form) representative of two-way action.

Two-way systems—especially those with multiple spans—often are supported on columns. As I just described, this type of support generally results in a high shear condition at the columns. One common means to relieve the interior force concentrations in the two-way spanning concrete flat slab is to extend the column's effective perimeter size by adding an enlarged top (called a *column capital*) and to increase the slab's strength locally by adding at the column a thick portion (called a *drop panel*).

In some truss systems you can develop something analogous to the drop panel in the concrete slab, which enables you to provide an additional trussing layer or a dropped-down unit to strengthen the truss.

Joints and System Assembly

When developing joints for spatial truss systems, you must consider many of the same issues you considered for planar truss systems: materials, truss member form, truss layout, and loads. You must focus on the basics before selecting jointing methods (welding, bolting, nailing, and so on) and deciding whether to use intermediate devices (gusset plates, nodal units, and so on).

For the spatial system two additional concerns are significant:

- Most joints are three-dimensional and thus require more complex joint form than the simple alternatives often possible for planar trusses.
- Such a system uses a lot of joints and thus requires a simple, economical joint construction.

Proprietary truss systems often involve a clever jointing system that accommodates a variety of joints, is inexpensive to mass-produce, and is quickly and easily assembled. *Note:* Although a joint system may correspond to a truss member shape, member selection usually is not as critical a design problem as the joint development.

The primary jointing methods in steel frameworks are

- Welding—member to member or member to primary joint element (gusset, node, and so on)
- Bolting—most likely to a joint element of some form
- Direct connection (e.g., threaded or snap-in)

Often, large structural units are prepared in a fabricating shop and then transported to the building site, where they are bolted together and to their supports. To avoid erection problems, carefully design field joints and monitor the shapes and sizes of shop-fabricated units.

As with all truss systems, joints are multifaceted. They serve many functions. They may accommodate thermal expansion, seismic separation, or some specific controlled structural action (for example, pinned joint response to avoid transfer of bending moments). They may support suspended elements.

A special joint occurs at a support for the truss system. Such a joint must be able to handle not only direct compression, but also tension or lateral forces (from wind or seismic actions) or both. In addition, such a joint may need to achieve some real pin joint re-

sponse to avoid transfer of bending to column tops as the truss
deflects. The design must factor in varying chord lengths (due to
thermal expansion or live load stress changes) and the composi-
tion of the supporting structure. See the discussion of *control joints*
in Section 11.8.

11

STEEL CONNECTIONS

A steel building structure typically consists of many parts. How you connect those parts depends on the parts' forms and sizes, the structural forces transmitted between parts, and the nature of the connecting materials. Today's primary connecting methods use electric arc welding and high-strength steel bolts. In this chapter I discuss such methods.

11.1 BASIC CONSIDERATIONS

Before designers choose a basic connecting method and designate the specific forms and sizes of connecting materials for an individual joint, they must consider several factors. Although designers should address these factors for every connection, practical reasons often limit the number of viable options.

Types of Connections

For steel parts, basic connection forms include the following:

Direct. If you directly join parts, you eliminate the need for intermediate elements. The threaded connection, which involves basic male and female coupling elements with cut spiral grooves, is a direct connection. Other examples are interlocking, folding, and direct bearing. Rarely are building structural elements directly connected.

Fusion. You can fuse separate metal pieces at their interface by welding, which is a popular structural connection.

Adhesion. You can use adhesive materials to form attachments—for example, through simple contact adhesion. Designers often develop composite units for walls, roofs, and floor panels with adhesion, but it is an uncommon structural connection for frameworks.

Intermediate Connector. You can use mechanical devices—including nails, staples, screws, rivets, and bolts—to connect parts. In some cases designers use a separate element, such as the angle pieces commonly used to develop beam connections. This connection type is often used for structural connections, especially with joints assembled at the job site.

Structural Functions

When designing a structural connection, you must take into account the forces transmitted between parts. For example, you must know if the connection faces a single loading condition or several loading conditions. Force actions are the following:

Direct Force. Tension, compression, and shear.

Bending. A rotational effect occurring in a plane that includes a member's linear axis. Connections usually transmit a single bending moment for any given loading condition; of course, a connection may face other bending moments for different loadings.

Torsion. A rotational effect occurring in a plane perpendicular to a member's linear axis. *Note:* Bending and torsion may occur together, as each produces different effects.

Reversible Forces. Opposite force effects—for example, compression versus tension (push/pull), back-and-forth shear, and clockwise versus counterclockwise bending or torsion. A connection facing reversible forces often needs some dual-resistance capability.

Dynamic Effects. Wind, earthquakes, mechanical vibrations, or intense sounds sometimes cause dynamic force actions on connections. By comparison, ordinary gravity effects are usually static. What works for static resistance may pop loose, develop metal fatigue, or otherwise fail under dynamic loading conditions. For example, from repeated dynamic effects, nuts may work off of bolts.

Designers must investigate every structural connection for the specific force types and magnitudes it must transmit. And because multiple loading is common, popular connections have a range of resistances.

Special Concerns

Connections must be economical and easy to achieve (even by unskilled laborers in some cases). Furthermore, the necessary materials must be available. Other concerns are the following:

Shop Versus Field. Connections performed under factory conditions (the *shop*) differ significantly from those performed at the job site (the *field*). Differences range from the basic means (bolting versus welding) to specific details. *Note:* In general, automation and quality control are more possible in the shop.

Permanence. Because assembled building structures usually stay in place indefinitely, designers rarely worry whether a connection is easy to undo. However, some structures are demountable, allowing part replacement or recycling. Bolts, screws, and interlocking connections permit disassembly more than welds, adhesives, rivets, and staples.

Movement-Selective. Most connections resist movements in the directions of applied forces, preventing separation, slipping, rotation, and so on. However, to accommodate thermal expansion, achieve seismic separation, or control selected structural responses, some joints may be selective, resisting some movements and facilitating others.

Connection design may relate to the structural type (i.e., truss, beam-and-column framework, rigid frame). Indeed, most steel frameworks employ common connecting means, using very familiar devices. For extensive structures, with many repetitive parts, designers may repeat connection types many times. The custom-designed connection is not only a design challenge, but it often requires too much effort, time, and money.

11.2   STEEL BOLTS

Designers often connect structural steel elements by mating flat parts with common holes and inserting a pin-type device. In times past the pin device was usually a rivet; today it is usually a bolt, available in many types and sizes. In this chapter I cover a few common bolting methods used in building structures.

Structural Actions of Bolted Connections

Figure 11.1 shows a simple connection between two steel bars. The connection functions to transfer a tension force from one bar to another. This tension-transfer connection is also known as a shear connection because of how the connecting device (the bolt) works. For structural connections, this joint type is achieved mostly with *high-strength bolts*—special bolts that are tightened in a way that induces yield stress in the bolt shaft. Connections using such bolts may fail in many ways, including the following:

Bolt Shear. In Figure 11.1, failure involves a slicing (shear) failure that is developed as a shear stress on the bolt cross section. The bolt resistance is equal to an allowable shear stress F_v times the area of the bolt cross section:

$$R = F_v \times A$$

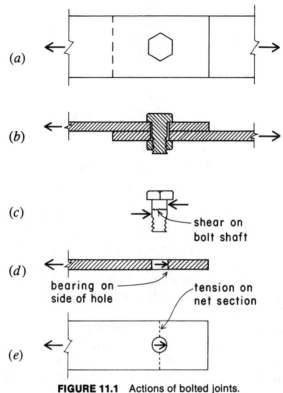

FIGURE 11.1 Actions of bolted joints.

If you know the bolt size and the steel grade, it is easy to establish this limit.

In some connections, you may need to slice the same bolt more than once to separate the connected parts; the bolt in Figure 11.2 must be sliced twice before the joint fails. When a bolt develops shear on only one section, it is said to be in *single shear*; when it develops shear on two sections, it is said to be in *double shear*.

Bearing. If the bolt tension (caused by a tightening of the nut) is relatively low, the bolt serves primarily as a pin in the matched holes, bearing against the sides of the holes (see

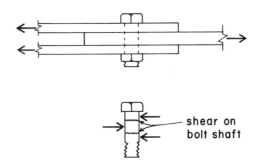

FIGURE 11.2 Bolt in double shear.

Figure 11.1*d*). When the bolt diameter is larger or the bolt is made of very strong steel, the connected parts must be sufficiently thick if they are to develop the bolt's full capacity. The maximum bearing stress (F_p) permitted for this situation by the AISC Specification is $1.2F_u$ where F_u is the ultimate tensile strength of the steel in the connected part in which the hole occurs.

Tension on Net Section of Connected Parts. For connected bars, the maximum tension stress occurs at the net section. Although the hole is a location of critical stress, you can achieve yield here before the connected parts experience serious deformation; for this reason, allowable stress at the net section is based on the bars' ultimate (rather than the yield) strength—normally $0.50F_u$.

Bolt Tension. Although the shear (slip-resisting) connection is common, some joints employ bolts for their resistance in tension (see Figure 11.3). For the threaded bolt, the maximum tension stress occurs at the net section through the cut threads. However, the bolt may elongate if yield stress develops in the bolt shaft (at an unreduced section). However you compute stress, bolt tension resistance is established from destructive tests.

Bending in the Connection. Whenever possible, designers use a bolted connection whose bolt layout is symmetrical with regard to the directly applied forces. When such a layout is impossible, the connection faces not only the direct force

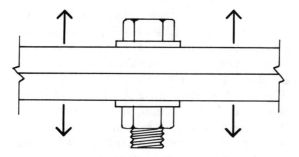

FIGURE 11.3 Tension on a bolt.

actions, but also twisting from a bending moment or torsion from the loads. Figure 11.4 shows some examples of this situation.

In Figure 11.4*a* two bars are connected by bolts. Because the bars are misaligned—that is, they cannot transmit ten-

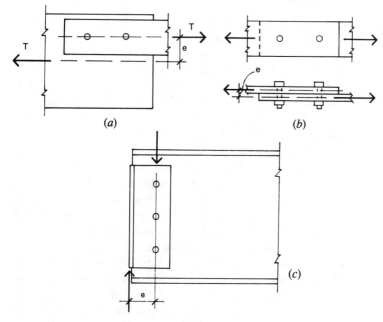

FIGURE 11.4 Bending in bolted joints.

sion directly between them—the bolts endure a rotational effect, with a torsional twist equal to the product of the tension force and the eccentricity due to the bars' misalignment. This twisting action increases the shear on the bolts. (Of course, the ends of the bars also twist.)

Figure 11.4b shows a single-shear joint, whose basic nature demands a twisting action. This twisting increases with thicker bars, so it is rarely critical for steel structures, whose connected elements are relatively thin. Avoid using this joint to connect wood elements, however.

Figure 11.4c is a side view of a beam end whose connection employs a pair of angles. As shown, the angles grasp the beam web between their legs and turn the other legs out to fit flat against a column or the web of another beam. Vertical load from the beam, vested in the shear in the beam web, is transferred to the angles by what connects the angles to the beam web—in this case, bolts. This load is then transferred from the angles at their outward-turned face; the eccentricity separates the forces.

Slipping of Connected Parts. Highly tensioned, high-strength bolts bring a very strong clamping action to the mated flat parts (see Figure 11.5). This action develops serious friction at the slip face. Friction is the initial form of resistance in the shear-type joint, preceding bolt shear, bearing, and even tension on the net section. For service-level loads, therefore, friction is the *usual* form of resistance, which is why the bolted joint with high strength (and highly tightened) bolts is known as a very rigid joint.

Block Shear. In a bolted connection, when the edge of an attached member is torn out, this failure is called *block shear*. In Figure 11.6, which shows a connection between two plates, such failure involves a combination of shear and tension. The total tearing force is the sum required to cause both forms of failure. The allowable stress on the net tension area is $0.50F_u$, where F_u is the maximum tensile strength of the steel. The allowable stress on the shear areas is $0.30F_u$. Once you know the edge distance, hole spacing, and hole diameters, determine the net widths for tension and shear. Then multiply these widths by the thickness of the part where the tear-

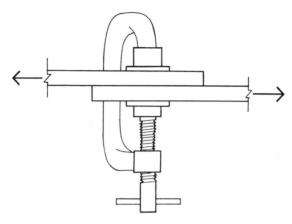

FIGURE 11.5   Slip-resisting clamping action of the high-strength bolt.

ing occurs. To find the total tearing force that can be resisted, multiply these areas by the appropriate stress. If this force is greater than the connection design load, the tearing problem is not critical.

Figure 11.7, another potential tearing case, is common: a beam is supported by another beam, and their tops align. You must cut back the end portion of the top flange of the supported beam to allow the beam web to extend to the side of the supporting beam. If you use a bolted connection, the tearing condition shown develops.

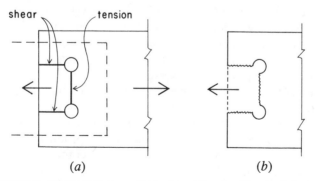

FIGURE 11.6   Tearing failure: a combination of shear and tension failures in the material connected by a bolted joint.

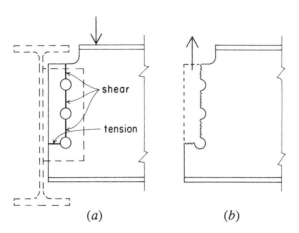

FIGURE 11.7 Tearing failure in a bolted beam connection.

The bottom line: You must investigate every bolted joint to identify particular critical conditions involved. Although the many variables make for a ton of possible outcomes, a few situations occur repeatedly.

Types of Steel Bolts

Bolts used to connect structural steel members come in two basic types:

- Bolts designated A307 are called *unfinished bolts* because they have the lowest load capacity of any structural bolts. The nuts for these bolts are tightened just enough so that attached parts attain a snug fit. Because of this low resistance to slipping, as well as the presence of oversized holes (to make assemblage more practical), there is some movement in the development of full resistance. In general, these bolts are not used for major connections, especially when joint movement or loosening under vibration or repeated loading is a problem. However, they are used extensively for temporary connections during frame erection.
- Bolts designated A325 or A490 are called *high-strength bolts*. The nuts for these bolts are tightened to produce a consider-

able tension force, which results in a high degree of friction resistance between attached parts. Different specifications for installation of these bolts results in different classifications of their strength, relating generally to the critical mode of failure.

When loaded in shear-type connections, a bolt's capacity is based on the shearing action in the connection. A bolt's shear capacity is further designated as S for single shear or D for double shear.

Table 11.1 lists the capacities of structural bolts ($\frac{3}{4}$ to $1\frac{1}{4}$ in. in diameter) in both tension and shear. Although bolts range in size from $\frac{5}{8}$ to $1\frac{1}{2}$ in. in diameter, the most common sizes for light structural steel framing are $\frac{3}{4}$ and $\frac{7}{8}$ in., as I describe in Section 11.3. For larger connections and large frameworks, common sizes are 1 to $1\frac{1}{4}$ in.

TABLE 11.1 Capacities of Structural Bolts (kips) [a]

ASTM Designation	Loading Condition [b]	Nominal Diameter of Bolt (in.)				
		$\frac{3}{4}$	$\frac{7}{8}$	1	$1\frac{1}{8}$	$1\frac{1}{4}$
		Area, Based on Nominal Diameter (in.^{2})				
		0.4418	0.6013	0.7854	0.9940	1.227
A307	S	4.4	6.0	7.9	9.9	12.3
	D	8.8	12.0	15.7	19.9	24.5
	T	8.8	12.0	15.7	19.9	24.5
A325	S	7.5	10.2	13.4	16.9	20.9
	D	15.0	20.4	26.7	33.8	41.7
	T	19.4	26.5	34.6	43.7	54.0
A490	S	9.3	12.6	16.5	20.9	25.8
	D	18.6	25.3	33.0	41.7	51.5
	T	23.9	32.5	42.4	53.7	66.3

[a] Slip-critical connections; assume that no bending exists in the connection and that bearing on connected materials is not critical.
[b] S = single shear; D = double shear; T = tension

Source: Adapted from data in the *Manual of Steel Construction*, 8th edition, with permission of the publisher, American Institute of Steel Construction.

Ordinarily, bolts are installed with a washer under both head and nut. A washer, however, is sometimes a limiting dimensional factor, precluding a bolt placement in some tight locations, such as close to the fillet (inside radius) of angles or other rolled shapes. Some manufactured high-strength bolts have specially formed heads or nuts that, in effect, have self-forming washers, eliminating the need for a separate washer.

For any given bolt diameter, you can determine a minimum thickness required for the bolted parts to develop the bolt's full shear capacity. This thickness is based on the bearing stress between the bolt and the side of the hole, which is limited to a maximum of $F_p = 1.5F_u$. The stress limit is established by either the bolt steel or the parts' steel.

Steel rods sometimes are threaded for use as anchor bolts or tie rods. When such rods are loaded in tension, their capacities are limited by the stress on the reduced section at the threads.

Tie rods sometimes are made with *upset ends*—that is, larger diameter portions at the ends. When these enlarged ends are threaded, the net section at the thread is the same as the gross section in the remainder of the rods. The result is no loss of capacity for the rod.

11.3   CONSIDERATIONS FOR BOLTED JOINTS

Layout of Bolted Connections

Designing bolted connections involves a number of factors regarding the dimensional layout of the bolt-hole patterns for the attached structural members. In this section I give an overview of the basic factors.

In the bolt pattern layout shown in Figure 11.8a, bolts sit in parallel rows. Two basic dimensions are limited by the bolt's size (nominal diameter).

- The *pitch* is the center-to-center spacing of the bolts. The AISC Specification limits this dimension to $2\frac{2}{3}$ times the bolt diameter. However, the preferred minimum, which I use in this book, is 3 times the diameter.

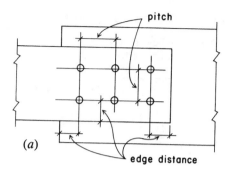

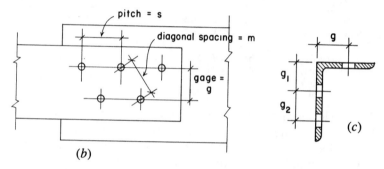

FIGURE 11.8 Layout considerations for bolted connections: (a) pitch and end distances (b) bolt spacing (c) gage distances for angle legs.

- The *edge distance* is the distance from the bolt's center to the nearest edge of the member containing the bolt hole. Naturally, edge distance is limited by bolt size and whether the edge is formed by rolling or is cut. Edge distance also may be limited by edge tearing in block shear, which I discussed in Section 11.2.

Table 11.2 gives the recommended limits for pitch and edge distance for the bolt sizes used in ordinary steel construction.

When bolts are staggered in parallel rows (Figure 11.8b), the diagonal distance, m, is important. For staggered bolts pitch is the bolt spacing in the direction of the rows, while the row spacing is called the *gage*. Designers stagger the bolts when they want a gage

TABLE 11.2 Pitch and Edge Distances for Bolts

Rivet or Bolt Diameter, d (in.)	Minimum Edge Distance for Punched, Reamed, or Drilled Holes (in.)		Minimum Recommended Pitch, Center to Center (in.)	
	At Sheared Edges	At Rolled Edges of Plates, Shapes, or Bars, or Gas-Cut Edges[a]	$2.667d$	$3d$
0.625	1.125	0.875	1.67	1.875
0.750	1.25	1.0	2.0	2.25
0.875	1.5^b	1.125	2.33	2.625
1.000	1.75^b	1.25	2.67	3.0

[a] May be reduced 0.125 in. when the hole is situated where stress does not exceed 25 percent of the maximum allowed in the connected element.
[b] May be 1.25 in. at the ends of beam connection angles.

Source: Adapted from data in the *Manual of Steel Construction*, 8th edition, with permission of the publisher, American Institute of Steel Construction.

spacing less than the pitch. In addition, staggering bolt holes also helps create a slightly less critical net section for tension stress.

The location of a bolt line often relates to the size and type of structural members you want to attach—especially when you plan to place bolts in the legs of angles or in the flanges of W, M, S, C, and structural tee shapes. For example, Figure 11.8c shows the legs of angles. When a single row is placed in a leg, it should sit at the distance g from the back of the angle. If you use two rows, the first row sits at the distance g_1 and the second row sits a distance g_2 from the first row. Table 11.3 gives the recommended values for g, g_1, and g_2.

TABLE 11.3 Usual Gage Dimensions for Angles (in.)

Gage Dimension	Width of Angle Leg								
	8	7	6	5	4	3.5	3	2.5	2
g	4.5	4.0	3.5	3.0	2.5	2.0	1.75	1.375	1.125
g_1	3.0	2.5	2.25	2.0					
g_2	3.0	3.0	2.5	1.75					

Source: Adapted from data in the *Manual of Steel Construction*, 8th edition, with permission of the publisher, American Institute of Steel Construction.

When placed at the recommended locations in rolled shapes, bolts end up a certain distance from the part's edge. From the recommended edge distances for rolled edges given in Table 11.2, you can determine the bolt's maximum size. For angles, the maximum fastener sometimes is limited by the edge distance, especially when two rows are used, but other factors may be more critical. For example, the distance from bolt center to the angle's inside fillet may prohibit the use of a large washer where one is required. Another factor is the stress on the angle's net section, especially when the member load is taken entirely by the attached leg.

Tension Connections

For tension members with reduced cross sections, you must perform two stress investigations. Such members include members with bolt holes or cut threads.

For members with holes, the allowable tension stress at the reduced cross section through the hole is $0.50F_u$, where F_u is the steels' ultimate tensile strength. You must compare the total resistance at this reduced section (also called the *net section*) with the resistance at other, unreduced sections (at which the allowable stress is $0.60F_y$).

For steel bolts the allowable tension stress is based on the bolt type. Table 11.1 gives the tension load capacities of three bolt types.

For threaded steel rods the allowable tension stress at the threads is $0.33F_u$.

When tension elements consist of W, M, S, and tee shapes, the tension connection usually precludes the attachment of all the parts of the section (e.g., both flanges plus the web for a W). In such cases, according to the AISC Specification, you must calculate a reduced effective net area A_e:

$$A_e = C_1 A_n$$

where A_n = the member's actual net area

C_1 = reduction coefficient

Unless you can justify a larger coefficient (through testing), use the following values:

- For W, M, or S shapes with flange widths not less than two-thirds the depth and structural tees cut from such shapes, when the connection is to the flanges and has at least three fasteners per line in the direction of stress, $C_1 = 0.90$.
- For W, M, or S shapes not meeting the above conditions and for tees cut from such shapes, provided the connection has not fewer than three fasteners per line in the direction of stress, $C_1 = 0.85$.
- For all members with connections that have only two fasteners per line in the direction of stress, $C_1 = 0.75$.

Angles used as tension members often are connected by only one leg. Conservative designers, use only the effective net area of the connected leg, less the area of the bolt holes. *Note:* Not only are rivet and bolt holes larger than the fastener's nominal diameter, but the punching damages a small amount of the steel around the hole's perimeter. Consequently, for design purposes, the hole diameter is always $\frac{1}{8}$ in. greater than the fastener's nominal diameter.

When only one hole is involved, or when you use a single row of fasteners along the line of stress, the net area of the plate cross section equals the product of the plate thickness and the plate's net width (i.e., member width minus hole diameter).

When holes are staggered in two rows along the line of stress (see Figure 11.9), the net section is determined somewhat differently. The AISC Specification reads:

> In the case of a chain of holes extending across a part in any diagonal or zigzag line, the net width of the part shall be obtained by deducting from the gross width the sum of the diameters of all the holes in the chain and adding, for each gage space in the chain, the quantity $s^2/4g$, where
>
> s = longitudinal spacing (pitch) in inches for any two successive holes.

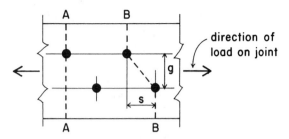

FIGURE 11.9   Determination of net cross-sectional area for connected members in a bolted connection.

and

g = transverse spacing (gage) in inches for the same two holes.

The critical net section of the part is obtained from that chain which gives the least net width.

The AISC Specification also states that the net section through a hole shall never be more than 85 percent of the corresponding gross section.

11.4   DESIGN OF A BOLTED CONNECTION

I illustrate the issues raised in the preceding sections in the following design example. Keep in mind the following:

- If slip-critical bolts are used, clean and smooth the surfaces of the connected members.
- If high-strength bolts are used, determine the particular ASTM specification for the bolt identity.
- Heed the following AISC Specification (Ref. 3) requirements:
  - Use at least two bolts per connection.
  - By the ASD method, the connection must carry a minimum load of 6 kips.

- Trusses, connections must develop at least 50 percent of the capacity of the connected members.

Note: Real-world design problems sometimes require designers to choose the fastener type and calculate the strengths required for the connected members. In the following problem, however, these are given.

Example 1. The connection in Figure 11.10 consists of a pair of narrow plates that transfer a tension force of 100 kips [445 kN] to a 10-in.-wide [250 mm] plate. All plates are of A36 steel with $F_y = 36$ ksi [250 MPa] and $F_u = 58$ ksi [400 MPa] and are attached with $\frac{3}{4}$ in. A325 bolts placed in two rows. Using data from Tables 11.1 and 11.2, determine the number of bolts required, the width and thickness of the narrow plates, the thickness of the wide plate, and the layout for the connection.

Solution: From Table 11.1, the capacity of a single bolt in double shear is 15.5 kips [69 kN]. The number of bolts required for the connection is thus

$$n = \frac{100}{15.5} = 6.45 \text{ or } 7$$

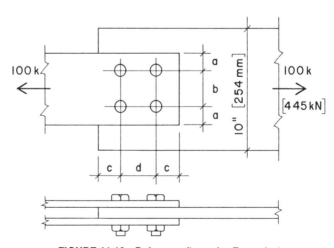

FIGURE 11.10 Reference figure for Example 1.

Although it is possible to place seven bolts in the connection, most designers prefer symmetrical arrangements (in this case, eight bolts, four to a row). The average bolt load is

$$P = \frac{100}{8} = 12.5 \text{ kips} \ [55.6 \text{ kN}]$$

From Table 11.2, for the $\frac{3}{4}$ in. bolts, minimum edge distance for a cut edge is 1.25 in. and minimum recommended spacing is 2.25 in. The minimum required plate width is thus

$$w = b + 2(a) = 2.25 + 2(1.25) = 4.75 \text{ in.} \ [121 \text{ mm}]$$

If space is tight, you can specify this width. For this example I use a width of 6 in.

The allowable stress on the gross area of the plate cross section is $0.60F_y = 0.60(36) = 21.6$ ksi. The required area is thus

$$A = \frac{100}{21.6} = 4.63 \text{ in.}^2 \ [2987 \text{ mm}^2]$$

Given the 6 in. width, the required thickness is

$$t = \frac{4.63}{2 \times 6} = 0.386 \text{ in.} \ [9.8 \text{ mm}]$$

As a result, you can use a minimum thickness of $\frac{7}{16}$ in. (0.4375) [11 mm]. Next check the stress on the net section, where the allowable stress is $0.50F_u = 0.50(58) = 29$ ksi [200 MPa]. For computations, use a bolt hole size at least $\frac{1}{8}$-in. larger than the bolt diameter to allow for the hole's true size and some steel damage at the hole edges. If you thus assume the hole diameter to be $\frac{7}{8}$ in. (0.875), the net width is

$$w = 6 - 2(0.875) = 4.25 \text{ in.} \ [108 \text{ mm}]$$

and the stress on the net section is

$$f_t = \frac{100}{2(0.4375 \times 4.25)} = 26.9 \text{ ksi} \ [187 \text{ MPa}]$$

This stress is lower than the allowable stress, so the narrow plates are adequate for tension stress.

The bolt capacities in Table 11.1 are based on a slip-critical condition, which assumes that the design failure limit is the bolts' friction resistance (slip resistance). The back-up failure mode is where the plates slip to permit the bolts' pin-action against the sides of the holes; this failure involves the bolts' shear capacity and the plates' bearing resistance. Because bolt capacities are always higher than the slip failures, the only concern is the bearing on the plates: according to the AISC Specification, $F_p = 1.2F_u = 1.2(58) = 69.6$ ksi [480 MPa].

To compute bearing stress, divide the load for a single bolt by the product of the bolt diameter and the plate thickness. Thus for the narrow plates,

$$f_p = \frac{12.5}{2 \times 0.75 \times 0.4375} = 19.05 \text{ ksi } [146 \text{ MPa}]$$

which is clearly not critical.

The procedure is essentially the same for the middle plate, except that the width is given and you are calculating for a single plate. As before, the stress on the unreduced cross section requires an area of 4.63 in.², so the required thickness of the 10-in.-wide plate is

$$t = \frac{4.63}{10} = 0.463 \text{ in. } [11.6 \text{ mm}]$$

In other words, the plate must be $\frac{1}{2}$ in. thick.

For this plate the width at the net section is

$$w = 10 - (2 \times 0.875) = 8.25 \text{ in. } [210 \text{ mm}]$$

and the stress on the net section is

$$f_t = \frac{100}{8.25 \times 0.5} = 24.24 \text{ ksi } [177 \text{ MPa}]$$

which is less than the allowable 29 ksi, as determined previously.

The computed bearing stress on the sides of the holes in the middle plate is

$$f_p = \frac{12.5}{0.75 \times 0.50} = 33.3 \text{ ksi } [243 \text{ MPa}]$$

which is less than the allowable 69.6 ksi, as determined previously.

The AISC Specification requires that the minimum spacing in the direction of the load be

$$\frac{2P}{F_u t} + \frac{D}{2}$$

and the minimum edge distance in the direction of the load be

$$\frac{2P}{F_u t}$$

where D = bolt diameter
  P = force transmitted by one bolt to the connected part
  t = thickness of the connected part

For this example, for the middle plate, the minimum edge distance is

$$\frac{2P}{F_u t} = \frac{2 \times 12.5}{58 \times 0.5} = 0.862 \text{ in.}$$

which is not critical. (The distance listed in Table 11.2 for the $\frac{3}{4}$ in. bolt at a cut edge is 1.25 in.)

For the minimum spacing

$$\frac{2P}{F_u t} + \frac{D}{2} = 0.862 + 0.375 = 1.237 \text{ in.}$$

which also is not critical.

A final problem: Calculate whether the two bolts at the end of a plate may be torn out in a block shear failure. Because the combined thicknesses of the outer plates is greater than that of the middle plate, the middle plate is critical. Figure 11.11 shows the tearing condition, which involves tension on section 1 and shear on the sections 2. For the tension section

$$\text{net } w = 3 - 0.875 = 2.125 \text{ in. } [54 \text{ mm}]$$

and the allowable stress for tension is

$$F_t = 0.50F_u = 29 \text{ ksi } [200 \text{ MPa}]$$

For the two shear sections

$$\text{net } w = 2\left(1.25 - \frac{0.875}{2}\right) = 1.625 \text{ in. } [41.3 \text{ mm}]$$

and the allowable stress for shear is

$$F_v = 0.30F_u = 17.4 \text{ ksi } [120 \text{ MPa}]$$

The total resistance to tearing is thus

$$T = (2.125 \times 0.5 \times 29) + (1.625 \times 0.5 \times 17.4)$$

$$= 44.95 \text{ kips } [205 \text{ kN}]$$

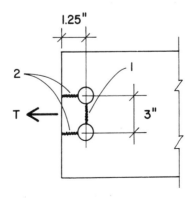

FIGURE 11.11   Tearing in Example 1.

Because this resistance is greater than the combined load on the two end bolts (25 kips), the plate is not critical for tearing in block shear.

Figure 11.12 is the complete connection solution.

Connections that transfer compression between joined parts are essentially the same regarding bolt stresses and bearing on the parts. Stress on the net section in the joined parts is rarely critical because the compression members usually are designed for a relatively low stress due to column action.

Problem 11.4.A. Use a bolted connection similar to that shown in Figure 11.10 to transmit a tension force of 175 kips [780 kN] by using $\frac{7}{8}$ in. A325 bolts and A36 steel plates. The outer plates are 8 in.

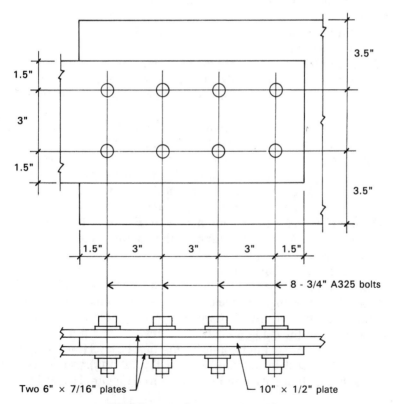

FIGURE 11.12 Solution for Example 1.

wide [200 mm] and the center plate is 12 in. wide [300 mm]. Find the required plate thicknesses. Also find the number of bolts needed if the bolts are placed in two rows. Sketch the connection layout.

Problem 11.4.B. Design a connection given the data in Problem 11.4.A, except the outer plates are 9 in. wide and the bolts are placed in three rows.

11.5 BOLTED FRAMING CONNECTIONS

How structural steel members are joined depends on the connected parts, the connecting device, and the forces that must transfer between members. Figure 11.13 shows several common connections used to join steel columns and beams made of rolled shapes.

In the joint shown in Figure 11.13a, a steel beam rests atop a steel plate that is welded to the top of a column. The bolts carry no loads if the force transfer is limited to the beam's vertical end reaction. The only stress condition of concern is a potential crippling of the beam web (see Section 6.9). In this situation you can use unfinished bolts.

The other Figure 11.13 details illustrate situations in which the beam reactions are transferred to the supports through the beam web. In general, this form of force transfer is appropriate because the vertical shear at the beam's end is resisted primarily by the beam web. The most common connection form uses a pair of angles—for example, to affix a steel beam to the side of a column (Figure 11.13b), the side of another beam (Figure 11.13d), or even the web of a W shaped column if the column depth provides enough space.

Figure 11.13c shows an alternative connection type: a single angle is welded to the side of a column, and the beam web is bolted to a side of the angle. Because the one-sided connection experiences some torsion, this form is generally acceptable only when the magnitude of the load on the beam is low.

When the tops of intersecting beams must be level (for example, to simplify a deck installation on top of the framing), you must cut back the top flange of the supported beam (see Figure 11.13e). Even

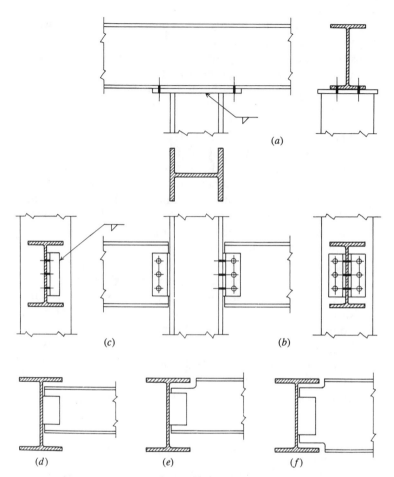

FIGURE 11.13   Typical bolted framing connections for light steel structures with rolled shapes.

worse is when the two beams have the same depth: then you must cut both flanges of the supported beam (see Figure 11.9*f*). Avoid these connection forms if possible, because they cost more and they reduces the beam's shear capacity. If either connection leads to critical shear in the beam web, you must reinforce the beam end. However, if the design calls for a steel deck, you may be able to adopt some form of the connection shown in Figure 11.14, which permits the beam tops to be offset by the depth of the deck ribs. Unless the flange of the supporting beam is quite thick, it should provide sufficient space to permit the connection, which does not require cutting the flange of the supported beam.

Figure 11.15 shows special framing details:

- The connections in Figure 11.15a are sometimes used when the supported beam is shallow. The vertical load is transferred through the seat angle, which may be bolted or welded to the carrying beam. The connection to the web of the supported beam provides additional resistance to roll over, or torsional rotation. Another advantage is the simplified field work: the seat angle may be welded in the shop, and the web connection requires only small unfinished bolts.
- Figure 11.15b shows a similar connection, used for joining a beam and column. For heavy beam loads the seat angle may be braced with a stiffening plate. A variation of this detail: if more than four bolts are required for attachment to the column, you may use two plates rather than the angle.
- Figures 11.15c and d show connections commonly used when pipe or tube columns carry the beams. Because the one-sided connection in Figure 11.15c produces some torsion in the beam, use the seat connection when the beam load is high.

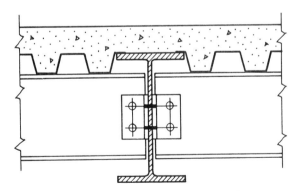

FIGURE 11.14 Special construction with deck-supporting beams; simplifies beam-to-girder connections.

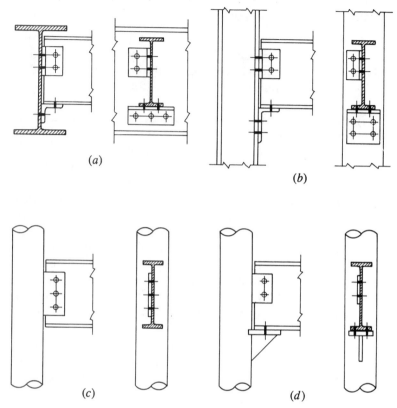

FIGURE 11.15   Bolted connections for special situations.

Framing connections often involve welding and bolting in a single connection. In general, welding is favored for shop work and bolting for field work. When developing connections, therefore, designers must understand the overall fabrication and erection process; they must be aware of what will be done where. Of course, the contractor inevitably will have some ideas about these procedures; in fact, the contractor may suggest alternate details, even for the best designs.

Connection details are particularly critical for structures containing a great deal of connections—for example, the truss.

Framed Beam Connections

The connection shown in Figure 11.13*b* is the type used most frequently in framed structures that consist of I shaped beams and H shaped columns. When developing this device, which is referred to as a *framed beam connection*, consider the following:

Type of Fastening. You can fasten angles to the supported beam and to the support with welds or bolts. The most common practice is to weld the angles to the supported beam's web in the shop and to bolt the angles to the support (i.e., the column face or the supporting beam's web) in the field.

Number of Fasteners. The number of bolts used on the supported beam's web. Although the outstanding legs of the angles require twice as many bolts as in the supported beam's web, the capacities are matched because the web bolts are in double shear. For smaller beams, or for light loads, angle leg sizes are typically just wide enough to accommodate a single row of bolts, as shown in Figure 11.16*b*. For very large beams, or for greater loads, use a wider leg to accommodate two rows of bolts.

Size of the Angles. Angle leg width and thickness depend on the size of fasteners and the magnitude of loads. Width of the outstanding legs also may depend on space available, especially if attachment is to the web of a column.

Length of the Angles. Angle length must accommodate the number of bolts. Figure 11.16 shows a standard layout, with bolts as large as 1 in. in diameter at 3 in spacing and end distance of 1.25 in. Angle length also depends on the distance available on the beam web—that is, the total length of the flat portion of the beam web (see Figure 11.16*a*).

The AISC Manual (Ref. 3) has information to support the design of frequently used connecting elements; data is provided for both bolted and welded fastenings. Most useful, predesigned connections are tabulated, so you can match them to loading magnitudes and to beam sizes.

Although there is no specified limit for the minimum size framed connection with a given beam, a general rule is to use one

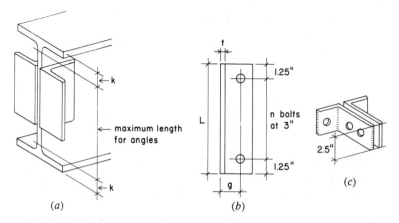

FIGURE 11.16 Framed beam connections for rolled shapes, using intermediate connecting angles.

whose angle length is at least one-half the beam depth. This rule helps ensure some minimum stability against rotational effects at the beam ends (see Sections 6.5 and 6.8).

For very shallow beams you may use the special connector shown in Figure 11.16c. Regardless of loading, this connector requires the angle leg at the beam web to accommodate two rows of bolts (one bolt in each row) simply for angle stability.

Of special concern is the inevitable bending in the connection (see Figure 11.4c). The bending moment arm for this twisting action is the gage distance of the angle leg—dimension g in Figure 11.16b. This bending is a reason to choose a relatively narrow angle leg.

If the top flange of the supported beam is cut back—as is common when connection is to another beam—vertical shear in the net cross section or block shear failure may be critical. Both conditions are aggravated when the supported beam has a very thin web, which is a frequent case because the most efficient beam shapes are usually the lightest shapes in their nominal size categories.

Another concern for the thin beam web: possible critical bearing stress in the bolted connection. As a result, try not to combine a large bolt with a beam whose web is thin.

Special connections usually are designed by the general structural designer. The framed beam connections I discussed here come from the AISC tabulations or are designed by employees of steel fabricators and erectors.

Bolted Truss Connections

A major element in truss design is the development of truss joints. Since a truss typically has several joints, the joints must be easy and economical to produce, especially if the building structural system uses a lot of similar trusses. When designing the joints, you must consider the truss configuration, member shapes and sizes, and the fastening method.

In most cases welding is the preferred fastening method for connections made in the shop, while bolting is used mostly for connections made at the building site. If possible, trusses are built in the shop; as a result, bolting is reserved for connections to supports, supported elements, and bracing, as well as splice points between shop-fabricated units. Of course, all is subject to change, depending on the rest of the building structure, the particular site, and the practices of local fabricators and erectors.

Figure 11.17 includes two common forms for light steel trusses. In Figure 11.17a the truss members consist of pairs of angles, and the joints are achieved with steel gusset plates (to which the members are attached). For top and bottom chords, the angles often are made continuous through the joint, reducing both the number of connectors required and the number of separate cut pieces of the angles. For some modest-size, flat-profiled, parallel-chorded trusses, the chords are made from tees, with interior members fastened directly to the tee web (see Figure 11.17b).

Figure 11.18 shows a layout for several joints of a light truss used for roofs with high slopes.

Developing the joint designs for the truss in Figure 11.18 involves many considerations, including

> *Truss Member Size, Load Magnitude.* Determines the size and type of connector required; the choice depends on individual connector capacity.
>
> *Angle Leg Size.* Relates to the bolt size; the choice depends on angle gages and minimum edge distances (see Table 11.3).

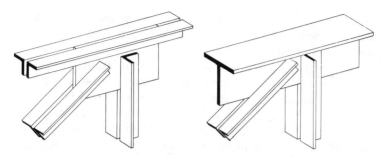

FIGURE 11.17 Common framing details for light steel trusses: (a) with chords consisting of double-angles and joints using gusset plates (b) with chords consisting of structural tees.

Thickness and Profile Size of Gusset Plates. Affect cost; naturally, designers prefer the lightest plates.

Layout of Members at Joints. To avoid twisting in the joint, designers want the action lines of the forces (vested in the rows of bolts) to meet at a single point.

Minimum edge distances for bolts (Table 11.2) can be matched to usual gage dimensions for angles (Table 11.3). And forces in members can be related to bolt capacities (Table 11.1)—if you keep the number of bolts to a minimum, the gusset plate is smaller.

Other issues require designers to manipulate the joint details. For really tight or complex joints, you may need to study the form of

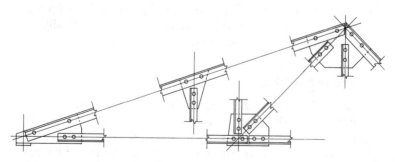

FIGURE 11.18 Typical light steel truss with double-angle truss members and bolted joints with gusset plates.

the joint with carefully drawn large-scale layouts. You can derive actual dimensions and forms (e.g., of member ends and gusset plates) from these drawings.

The truss in Figure 11.18 has some common features. For example, all member ends are connected by only two bolts, the minimum required, which means that the minimum-size bolt chosen has sufficient capacity to develop the forces in all members. Another example: At the top chord joint between the support and the peak, the top chord member is continuous (uncut) at the joint, a common and cost-effective case where the available member lengths are greater than the joint-to-joint distances in the truss.

If a building needs only a few trusses, the fabrication may be as shown in Figure 11.18.

11.6 WELDING

In some instances welding is an alternative to bolting. Most often, a connecting device (bearing plate, framing angles, etc.) is welded to one member in the shop and bolted to a connecting member in the field. However, in many instances, joints are fully welded, whether done in the shop or on site. For some situations welding is the only reasonable means to make a joint. In the following sections I present some of the problems and potential uses of welding in building structures.

One advantage of welding is that you can directly connect members, eliminating the need for intermediate devices such as gusset plates or framing angles. Another advantage: no holes, so you can maximize the capacity of the unreduced cross section of tension members. Welding also enables you to develop exceptionally rigid joints, an advantage in moment-resistive connections.

Electric Arc Welding

Electric arc welding is the process generally used in steel building construction. In this type of welding, an electric arc is formed between an electrode and the two pieces of metal you want to join. The globules of melted metal from the electrode flow into the molten seat; when cool, they are united with the members you wanted to weld together.

The term *penetration* indicates the depth from the original surface of the base metal to the point at which fusion ceases. *Partial penetration,* the failure of the weld metal and base metal to fuse at the root of a weld, produces inferior welds.

Welded Connections

Several joints are shown in Figure 11.19. In general, there are three classifications of joints: *butt joints, tee joints,* and *lap joints.* Which weld you use depends on the magnitude of the load requirement, the manner in which it is applied, and your budget. A detailed discussion of the many joints and their uses and limitations is beyond the scope of this book.

A weld commonly used for structural steel in building construction is the *fillet* weld. Approximately triangular in cross section, it is formed between the two intersecting surfaces of the joined members (see Figures 11.20a and b). The *size* of a fillet weld is the leg length of the largest inscribed isosceles right triangle, AB or BC (see Figure 11.20a). The *throat* of a fillet weld is the distance from the root to the hypotenuse of the largest isosceles right triangle that can be inscribed within the weld cross section—distance BD in Figure 11.20a. The exposed surface of a weld is not the plane surface indicated in Figure 11.20a; it is usually somewhat convex, however, as shown in Figure 11.20b. Therefore the actual throat may be greater than that shown in Figure 11.20a, but do not consider this additional material, called *reinforcement*, when determining a weld's strength.

Stresses in Fillet Welds

If *AB* is one unit,

$$(AD)^2 + (BD)^2 = (1)^2$$

Because *AD* and *BD* are equal,

$$2 (BD)^2 = (1)^2$$

or

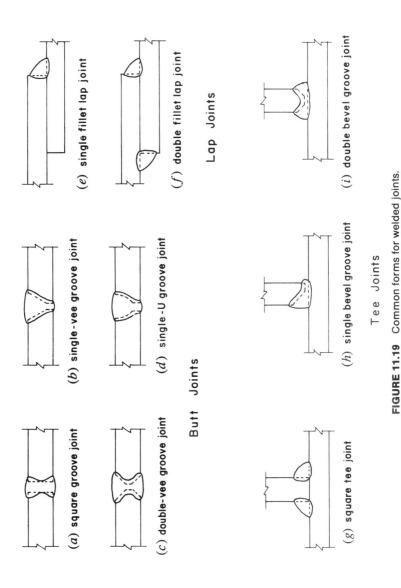

(a) square groove joint

(b) single-vee groove joint

(c) double-vee groove joint

(d) single-U groove joint

Butt Joints

(e) single fillet lap joint

(f) double fillet lap joint

Lap Joints

(g) square tee joint

(h) single bevel groove joint

(i) double bevel groove joint

Tee Joints

FIGURE 11.19 Common forms for welded joints.

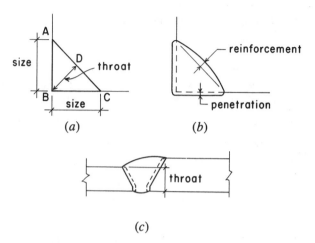

FIGURE 11.20 Dimensional considerations for welds.

$$(BD)^2 = \frac{1}{2}$$

$$BD = \sqrt{0.5} = 0.707$$

In other words, the throat of a fillet weld is equal to the *size* of the weld multiplied by 0.707. Consider a $\frac{1}{2}$ in. fillet weld—that is, a weld with dimensions AB or BC equal to $\frac{1}{2}$ in. The weld's throat is 0.5 × 0.707, or 0.3535 in. If the allowable unit shearing stress on the throat is 21 ksi, the allowable working strength of a $\frac{1}{2}$ in. fillet weld is 0.3535 × 21 = 7.42 kips/in. of weld. If the allowable unit stress is 18 ksi, the allowable working strength is 0.3535 × 18 = 6.36 kips/in. of weld length.

The permissible unit stresses discovered in the preceding paragraph are for welds made with E 70 XX- and E 60 XX-type electrodes on A36 steel. Note that the stress in a fillet weld is considered as shear on the throat, regardless of the direction of the applied load. Table 11.4 lists the allowable working strengths of fillet welds of various sizes.

The stresses allowed for the metal of the connected parts (known as the *base metal*) apply to complete penetration groove welds that

TABLE 11.4   Safe Service Loads for Fillet Welds

Size of Weld (in.)	Allowable Load (kips/in.)		Allowable Load (kips/in.)		Size of Weld (mm)
	E 60 XX Electrodes	E 70 XX Electrodes	E 60 XX Electrodes	E 70 XX Electrodes	
$\frac{3}{16}$	2.4	2.8	0.42	0.49	4.76
$\frac{1}{4}$	3.2	3.7	0.56	0.65	6.35
$\frac{5}{16}$	4.0	4.6	0.70	0.81	7.94
$\frac{3}{8}$	4.8	5.6	0.84	0.98	9.52
$\frac{1}{2}$	6.4	7.4	1.12	1.30	12.7
$\frac{5}{8}$	8.0	9.3	1.40	1.63	15.9
$\frac{3}{4}$	9.5	11.1	1.66	1.94	19.1

are stressed in tension or compression parallel to the weld axis or in tension perpendicular to the effective throat. They also apply to complete or partial penetration groove welds stressed in compression normal to the effective throat and in shear on the effective throat. Consequently, allowable stresses for butt welds are the same as for the base metal.

Table 11.5 shows the relationship between the weld size and the maximum thickness of material in joints connected only by fillet welds. The maximum size of a fillet weld applied to the square edge of a plate or section that is $\frac{1}{4}$ in. or more in thickness should be $\frac{1}{16}$ in. less than the nominal thickness of the edge. Along edges of material less than $\frac{1}{4}$ in. thick, the maximum size may be equal to the thickness of the material.

The effective area of butt and fillet welds is the product of the weld's effective length and the effective throat thickness. The minimum effective length of a fillet weld should not be less than four times the weld size. For starting and stopping the arc, add a distance approximately equal to the weld size to the design length of fillet welds when specifying to the welder.

Figure 11.21a represents two plates connected by fillet welds. The welds marked A are longitudinal; weld B is a transverse weld. If a load is applied in the direction shown by the arrow, the stress dis-

TABLE 11.5 Relation Between Material Thickness and Size of Fillet Welds

Material Thickness of the Thicker Part Joined		Minimum Sizes of Fillet Weld	
in.	mm	in.	mm
To $\frac{1}{4}$ inclusive	To 6.35 inclusive	$\frac{1}{8}$	3.18
Over $\frac{1}{4}$ to $\frac{1}{2}$	Over 6.35 to 12.7	$\frac{3}{16}$	4.76
Over $\frac{1}{2}$ to $\frac{3}{4}$	Over 12.7 to 19.1	$\frac{1}{4}$	6.35
Over $\frac{3}{4}$	Over 19.1	$\frac{5}{16}$	7.94

tribution in the longitudinal weld is not uniform, and the stress in the transverse weld is approximately 30 percent higher per unit of length.

A transverse fillet weld that terminates at the end of a member (see Figure 11.21b) gains strength if the weld is returned around the corner for a distance not less than twice the weld size. These end returns, sometimes called *boxing*, help resist tearing action on the weld.

The $\frac{1}{4}$ in. fillet weld is the smallest practical weld, and the $\frac{5}{16}$ in. weld is probably the most economical weld. A small continuous weld is generally more economical than a larger discontinuous weld if both are made in one pass. Some specifications limit the single-pass fillet weld to $\frac{5}{16}$ in. Large fillet welds require two or more passes (multipass welds), as shown in Figure 11.21c.

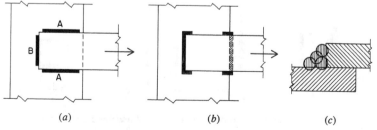

(a)   (b)   (c)

FIGURE 11.21 Welding of lapped steel elements.

11.7 DESIGN OF WELDED CONNECTIONS

Members must be held firmly in position during welding. Often temporary connection devices are necessary. In fact, in the field you may need a complete temporary erection connection, even though this connection is redundant after welding.

Although welding in the shop is often automated, field welding is almost always achieved by hand.

In the following example I demonstrate how to design simple fillet welds for ordinary connections.

Example 1. Weld a bar of A36 steel, $3 \times \frac{7}{16}$ in. [76.2 × 11 mm] in cross section, with E 70 XX electrodes to the back of a channel so that the bar's full tensile strength is developed. Determine the size of the required fillet weld. (See Figure 11.22.)

Solution: The usual allowable tension stress for this situation is $0.6F_y$; thus

$$F_a = 0.6(F_y) = 0.6(36) = 21.6 \text{ ksi}$$

and the tension capacity of the bar is

$$T = F_a A = 21.6(3 \times 0.4375) = 28.35 \text{ kips}$$

The weld must be large enough to resist this force.

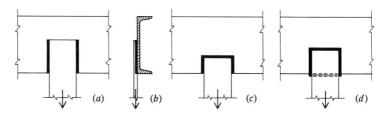

FIGURE 11.22 Variations of the form of a welded connection.

A practical weld size is $\frac{3}{8}$ in., for which Table 11.5 yields a strength of 5.6 kips/in. The required length to develop the bar strength is

$$L = \frac{28.35}{5.6} = 5.06 \text{ in.}$$

Adding a minimum distance equal to the weld size to each end for start and stop of the weld, a practical length is 6 in.

Figure 11.22 shows three ways to arrange the weld. In Figure 11.2*a*, the weld is divided into two equal parts. For two starts and stops, place 4 in. of weld on each side of the bar.

The weld in Figure 11.22*c* has three parts: a 3-in.-long weld across the end of the bar and a 3 in. weld you should split between the bar's two sides. Each weld is 2 in. long to ensure a total of 3 in. of effective weld.

Neither of the welds in Figures 11.22*a* or *c* resists the twisting action on the unsymmetrical joint well enough. To accommodate this action, most designers provide some additional weld. The best weld is shown in Figure 11.22*d*, where a weld is provided on the back of the bar, between the bar and the corner of the channel. You can develop this weld as an addition to either of the welds in Figures 11.22*a* or *c*. The weld on the back is primarily a stabilizing weld, and so you cannot count on it to resist the required tension force.

Developing a welded joint combines computations and judgments.

Example 2. Connect a $3\frac{1}{2} \times 3\frac{1}{2} \times \frac{5}{16}$-in. [89 × 89 × 8 mm] angle of A36 steel subjected to a tensile load to a plate by fillet welds, using E 70 XX electrodes. Find dimensions of the welds to develop the angle's full tensile strength.

Solution: From Table A.4 the cross section area of the angle is 2.09 in.^{2} [1348 mm^2]. The maximum allowable tension stress is $0.60F_y = 0.60(35) = 21.6$ ksi [150 MPa]. Thus the tensile capacity of the angle is

$$T = F_t A = (21.6)(2.09) = 45.1 \text{ kips [200 kN]}$$

For the $\frac{5}{16}$ in. angle leg thickness, the maximum recommended weld is $\frac{1}{4}$ in. From Table 10.5, the weld capacity is 3.7 kips/in. The total length of weld required is thus

$$L = \frac{45.1}{3.7} = 12.2 \text{ in. } [310 \text{ mm}]$$

You can divide this total length between the two sides of the angle, but, assuming the tension load in the angle to coincide with its centroid, the distribution of the load to the two sides is not equal. Thus some designers prefer to proportion the lengths of the two welds so that they correspond to their positions on the angle. To do so, they use the following procedure.

From Table A.4, the centroid of the angle is at 0.99 in. from the back of the angle. The two weld lengths shown in Figure 11.23 should be in inverse proportion to their distances from the centroid. Thus

$$L_1 = \frac{2.51}{3.5} \times 12.2 = 8.75 \text{ in. } [222 \text{ mm}]$$

and

$$L_2 = \frac{0.99}{3.5 \times 12.2} = 3.45 \text{ in. } [88 \text{ mm}]$$

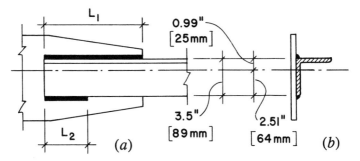

FIGURE 11.23 Form of the welded connection in Example 1.

However, make these required design lengths at least $\frac{1}{4}$ in. longer at each end. Thus, reasonable specified lengths are $L_1 = 9.25$ in., $L_2 = 4.0$ in.

When angle shapes are used as tension members, and when they are connected at their ends by fastening only one leg, do not assume a stress distribution of equal magnitude on the entire angle cross section. Some designers prefer to ignore the development of stress in the unconnected leg and to limit the member capacity to the force obtained by considering only the connected leg. In this example, the maximum tension is thus reduced to

$$T = F_t \times A = (21.6) \times (3.5 \times 0.3125)$$

$$= 23.625 \text{ kips } [105 \text{ kN}]$$

and the required total weld length is

$$L = \frac{23.625}{3.7} = 6.39 \text{ in. } [162 \text{ mm}]$$

Then divide this length evenly between the two sides. After adding an extra length of twice the weld size, a specified length is 3.75 in. on each side.

Problem 11.6.A. A $4 \times 4 \times \frac{1}{2}$-in. angle of A36 steel is to be welded to a plate with E 70 XX electrodes to develop the full tensile strength of the angle. Using $\frac{3}{8}$-in. fillet welds, compute the design lengths for the welds on the two sides of the angle, assuming development of tension on the full cross section of the angle.

Problem 11.6.B. Redesign the welded connection in Problem 11.6.A. assuming that the tension force is developed only in the connected leg of the angle.

Plug and Slot Welds

One method of connecting two overlapping plates uses a weld in a hole made in one of the two plates (see Figure 11.24). Plug and slot welds are those in which the entire area of the hole or slot receives weld metal. The maximum and minimum diameters of plug and slot welds and the maximum length of slot welds are shown in Figure 11.24. If the plate containing the hole is not more than $\frac{5}{8}$ in. thick, the hole should be filled with weld metal. If the plate is more than $\frac{5}{8}$ in. thick, the weld metal should be at least one-half the thickness of the material but not less than $\frac{5}{8}$ in.

The stress in a plug or slot weld is considered to be shear on the area of the weld at the plane of contact of the two plates being connected. The allowable unit shearing stress, when E 70 XX electrodes are used, is 21 ksi [145 MPa].

A somewhat similar weld consists of a continuous fillet weld at the circumference of a hole, as shown in Figure 11.24c. This is not a plug or slot weld and is subject to the usual requirements for fillet welds.

11.8 WELDED STEEL FRAMES

Welding is presently used extensively to assemble steel frames, both in the shop and in the field. But in the nineteenth century, most assembly was done with rivets. When bolting and welding gained prominence in the mid-twentieth century, riveting lost its

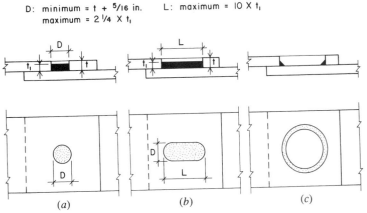

FIGURE 11.24 Welds placed in holes: (a) plug weld (b) slot weld (c) fillet weld in a large hole.

hold. Nevertheless, the classic joint forms developed for riveting are still widely used.

In many cases, joints may be developed with bolting or welding. The decision often has to do with location of the work—shop or field. Many of the joints shown as bolted in this book would work as well if welded.

In some cases joints are developed with some assembly achieved in the fabricating shop and some in the field. A common example is the standard framed beam connection using a pair of angles. It is now common to attach the angles to the supported beam's ends in the shop by welding and to complete the connection to the supporting member in the field with bolts.

A special structure is the multistory, multi-bay steel frame for multilevel buildings. Parts of these three-dimensional frameworks are frequently developed as rigid-frame bents to resist lateral forces due to wind or earthquakes. These bents have mostly been achieved in recent years by welding beams directly to the columns. However, recent experiences in major earthquakes have caused reconsideration of these jointing methods and new forms are being developed where seismic design is critical.

11.9 CONTROL JOINTS: DESIGN FOR SELECTED BEHAVIOR

Most connections between elements of a structure are designed to *resist* movement within the joint. The usual purpose of fastening devices, such as bolts and welds, is to hold the connected parts firmly together. A *control joint*, on the other hand, is a joint in which some form of movement is deliberately facilitated. The three most common reasons for such a joint are the following:

> *Thermal Expansion.* All materials expand and contract with changes in temperature. In large buildings, or any long building part, the total movement can be considerable. This requires some consideration for the forces that can accumulate and may break up the construction.
>
> *Structural Actions.* Actions of continuous and longspan structures can cause movements at their supports that can disrupt

other construction. The support joint may need to accommodate some aspects of this movement with reduced resistance, while providing the basic support that is mandatory.

Seismic Separation. Multi-massed buildings may have discrete parts that want to move differently under the dynamic effect of an earthquake. One technique for dealing with this is to provide a tolerance for independent movement of the parts at their points of connection.

Of course the simplest form of a control joint is no connection—that is, the non-joining of parts. From a structural point of view, this is essentially what most seismic separation joints are. However, a common situation is that in which some form of connection is required, while simultaneously facilitating some form of movement. For example, it may be desirable to support a beam in a manner that does not restrict its end rotation; in other words, provide a simple support for vertical force, not a fixed one.

A common device used for bolted connections is the slotted hole, as shown in Figure 11.25. If the width of the hole is only slightly larger than the bolt diameter, the plate is reasonably restrained from moving in that direction, even if the bolt is not tight.

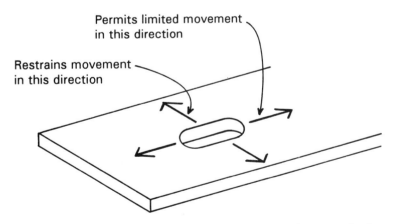

FIGURE 11.25 Commonly means for facilitating movement in a connection in a single direction. Also used to provide for anticipated lack of accuracy of placement of members in a framing system.

In the other direction, however, with the bolt in place, the plate can be moved a small distance. This device can be used to permit small movements (such as those from thermal expansion) but is actually more frequently used for tolerance of erection inaccuracies.

There are many types and many details for control joints. The following example illustrates a situation where some kind of structural resistance is required, but other resistances need to be reduced as much as possible.

The behavior of trusses requires some allowance for change in the length of the chords as loading changes—basically, as the live load comes and goes. This implies that the truss supports can facilitate some overall change in the length of the truss, which requires some change in the actual distance between the support points. If movements are small in actual dimension and there is some possibility of nondestructive deformation in the truss-to-support connections, it may not be necessary to make any special provision for the movements. If, however, the actual dimensions of movements are large, and both the support structure and the connections to it virtually unyielding, problems will occur unless actual provision is effectively made to facilitate the movements.

A wide range in temperature can also produce considerable length change in long structures. These effects should be considered in terms of movements at the supports as well as length change. Two examples:

- *Long-Span Steel Truss, Supported by Masonry Piers.* In this case the movements will be considerable and the supports essentially unyielding. Special provision must be made at one or both supports for some actual dimension of movement.
- *Truss Erected During Cold Weather But Later Subjected to Warm Weather or to the Warmer Conditions Maintained in the Enclosed Building.* In this case, even though provisions for movement due to loading stresses may be unnecessary, the length changes due to thermal change should be considered.

A technique that is sometimes used to reduce the need for any special provision for movement at supports is to leave the support connections in a stable but untightened condition until after the building construction is essentially completed. This allows the

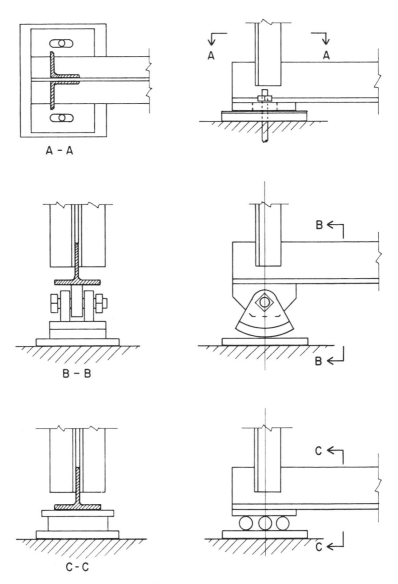

FIGURE 11.26 Support details for steel trusses; used to accommodate horizontal movement at the joint. (a) Sliding connection with a friction-reducing slip material between the steel bearing plates. (b) Pinned rocker. (c) Hardened steel roller bearings between the bearing plates.

truss deformations resulting from the dead load to accumulate during construction so that the critical effects are limited to the deformations caused by the live loads. Where the dead load is a major part of the total design load, this is often quite effective.

When provision must be made for movement, the precise form of movements and their approximate anticipated magnitudes must be carefully determined. For support of a horizontal-spanning structure, such as a beam or truss, the minimum structural resistance required is that in response to vertical force. Thus the joint must restrict vertical movement, while possibly tolerating both horizontal movement and rotation.

Figure 11.26 shows several details that may be used where provision for movement must be made at the truss supports. The need for these is a matter of judgment and must be considered in terms of the full development of the building construction.

In Figure 11.26a the method used is a slight modification of a common joint that uses an end bearing plate bolted to the support with anchor bolts embedded in masonry or concrete. In this version a second steel plate is used; thus one plate comes welded to the truss and the other is set in place on the top of the wall. Slotted holes (Figure 11.25) are provided in the truss end plate. These also provide for some inaccuracy in the precise positioning of the anchor bolts—a wise choice in any regard. However, if the nuts on the anchor bolts are only hand-tightened and a friction-reducing interface material is used, resistance to horizontal movement is minimal.

The method shown in Figure 11.26b uses a rather elaborate rocker device. An advantage gained with this joint is the very low resistance to rotation, if indeed this is a concern.

The detail in Figure 11.26c is a variation on a so-called *roller bearing joint*. In this case three steel rods are used to separate the truss end plate and the wall top plate. Horizontal movement is thus only slightly resisted. However, as in Figure 11.26a, there is little provision for reduction of effects of end rotation of the truss.

Since control joints are often provided because of possible effects on the surrounding or supported construction, their effectiveness must be evaluated for both structural adequacy and overall value to the building.

12

BUILDING STRUCTURES: DESIGN EXAMPLES

In this chapter I present several case examples. I chose these examples to illustrate as much as possible the concepts developed in the earlier chapters. For detailed discussion of individual structural elements, however, refer to those other chapters; here I focus on the development of whole building structural systems.

12.1 GENERAL CONSIDERATIONS FOR BUILDING STRUCTURES

Building construction materials, methods, and details vary considerably region by region for many reasons, including climate. Even in a single region, you can find differences between similar buildings, reflecting individual architectural design styles and building techniques. Nevertheless, at any given time, most buildings of a similar type and size employ a few predominant, popular construction methods. *Note:* In no way do I endorse the construction methods and details shown in this chapter as superior building styles.

Structural Planning

When planning a structure—whether the building is simple or complex, small or large, ordinary or unique—you must perform two major tasks:

- Logically arrange the structure (taking into account its geometric form, dimensions and proportions, and elements) for basic stability and reasonable interaction. Spanning beams must be supported and have depths adequate for the spans, horizontal thrusts of arches must be resolved, columns above must be centered over columns below, and so on.
- Develop the relationships between the structure and the building in general. In other words, "see" the structural plan inherent in the architectural plan. The structural plan and the building plan do not need to be identical, but they must fit together. Likewise architectural planning and structural planning should be done interactively, not separately. The more the architect knows about the structural problems and the more the structural designer knows about the architectural problems, the more likely an interactive design development will occur.

Although each building is unique, most building design problems are alike. Choosing a final design involves evaluating the many possible solutions. The more common the problem, the more well known the possible solutions. However, when the problem is new—for example, a new building use, a jump in scale, or a new performance situation—you need innovation. Fortunately, the combination of technology and imagination inevitably lead to new solutions, even for the most common problems. (Of course, do not choose a new solution to an old problem without comparing it to established solutions.) The bottom line: When selecting a solution, consider all possible alternatives: those well known, those new and unproven, and those only imagined.

Building Systems Integration

Good structural design integrates the structure into the building's whole physical system. You must realize how structural design decisions influence not only the architectural design but also the development of systems for power, lighting, thermal control, ventilation, water supply, waste handling, vertical transportation, firefighting, and so on. The most popular structural systems accommodate building subsystems *and* facilitate popular architectural forms and details.

12.2   DESIGN LOADS

As I mentioned in Chapter 5, structures exist essentially to resist loads. Thus designers must thoroughly understand every load's nature, source, and effects. In this section I offer a brief summary of major common loads.

Building Code Requirements

Structural design must adhere to local building codes, which are government-administered. If a design does not conform to the local code, the government unit in charge will not grant a building permit—that is, the legal permission required for construction.

   Most U.S. building codes are based on one of three model codes:

- *Uniform Building Code* (UBC) (Ref. 1), which is widely used in the west because it has the most complete data for seismic design.
- *BOCA Basic National Building Code*, used in the east and midwest.
- *Standard Building Code*, used in the southeast.

Largely derived from the same basic data and standard reference sources, these model codes are more similar than different. Local codes, however, reflect particular regional concerns.

   All codes cover the following issues:

Minimum Required Live Loads. All codes have tables similar to Tables 12.2 and 12.3.

Wind Loads. Depend on local windstorm conditions. Model

codes provide data broken down by geographic zone.

Seismic (Earthquake) Effects.   Depend on local conditions. This data, including recommended investigations, is frequently modified, reflecting ongoing research and experience.

Load Duration.   When calculating loads or design stresses, designers often must take into account the time span of the load, which varies from the life of the structure (for dead load) to a fraction of a second (for a wind gust or a seismic shock). I offer some examples later in this chapter.

Load Combinations.   Formerly left to the discretion of designers, combinations now are commonly spelled out in codes because more designers are using ultimate strength design and factored loads.

Design Data for Types of Structures.   All codes cover basic materials (wood, steel, concrete, masonry, etc.), specific structures (rigid frames, towers, balconies, pole structures, etc.), and special problems (foundations, retaining walls, stairs, etc.). Although local codes generally recognize industry-wide standards and common practices, they also tend to reflect local experience or attitudes. Most codes define minimal structural safety, so following specified limits may result in questionable performances (bouncy floors, cracked plaster, etc.).

Fire Resistance.   Codes are concerned with two primary effects: structural collapse and containment of the fire to control its spread. Designers must heed these concerns, even when code limits restrict your choice of materials or construction details.

Note: The work in this chapter's examples is based largely on UBC criteria because I am most familiar with its specifics.

Dead Load

Dead load is a material's weight—when designing a beam, you must allow for the weight of the beam itself. (Table 12.1 lists the weights of many construction materials.) Dead load is due to gravity and results in a downward vertical force. Dead load is a permanent load after the building construction is completed, unless the building is remodeled.

TABLE 12.1 Weights of Construction Materials

	psf[a]	kPa[a]
Roofs		
3-ply ready roofing (roll, composition)	1	0.05
3-ply felt and gravel	5.5	0.26
5-ply felt and gravel	6.5	0.31
Shingles: Wood	2	0.10
Asphalt	2–3	0.10–0.15
Clay tile	9–12	0.43–0.58
Concrete tile	6–10	0.29–0.48
Slate, $\frac{1}{4}$ in.	10	0.48
Insulation: Fiberglass batts	0.5	0.025
Foam plastic, rigid panels	1.5	0.075
Foamed concrete, mineral aggregate	2.5/in.	0.0047/mm
Wood rafters: 2 × 6 at 24 in.	1.0	0.05
2 × 8 at 24 in.	1.4	0.07
2 × 10 at 24 in.	1.7	0.08
2 × 12 at 24 in.	2.1	0.10
Steel deck, painted: 22 gage	1.6	0.08
20 gage	2.0	0.10
Skylights: Steel frame with glass	6–10	0.29–0.48
Aluminum frame with plastic	3–6	0.15–0.29
Plywood or softwood board sheathing	3.0/in.	0.0057/mm
Ceilings		
Suspended steel channels	1	0.05
Lath: Steel mesh	0.5	0.025
Gypsum board, $\frac{1}{2}$ in.	2	0.10
Fiber tile	1	0.05
Drywall, gypsum board, $\frac{1}{2}$ in.	2.5	0.12
Plaster: Gypsum	5	0.24
Cement	8.5	0.41
Suspended lighting and HVAC, average	3	0.15
Floors		
Hardwood, $\frac{1}{2}$ in.	2.5	0.12
Vinyl tile	1.5	0.07
Ceramic tile: $\frac{3}{4}$ in.	10	0.48
Thin-set	5	0.24
Fiberboard underlay, 0.625 in.	3	0.15
Carpet and pad, average	3	0.15
Timber deck	2.5/in.	0.0047/mm
Steel deck, stone concrete fill, average	35–40	1.68–1.92

(*Continued*)

TABLE 12.1 *Continued*

	psf [a]	kPa [a]
Concrete slab deck, stone aggregate	12.5/in.	0.024/mm
Lightweight concrete fill	8.0/in.	0.015/mm
Wood joists: 2 × 8 at 16 in.	2.1	0.10
2 × 10 at 16 in.	2.6	0.13
2 × 12 at 16 in.	3.2	0.16
Walls		
2 × 4 studs at 16 in., average	2	0.10
Steel studs at 16 in., average	4	0.20
Lath, plaster—see *Ceilings*		
Drywall, gypsum board, $\frac{1}{2}$ in.	2.5	0.10
Stucco, on paper and wire backup	10	0.48
Windows, average, frame+glazing:		
Small pane, wood or metal frame	5	0.24
Large pane, wood or metal frame	8	0.38
Increase for double glazing	2–3	0.10–0.15
Curtain wall, manufactured units	10–15	0.48–0.72
Brick veneer, 4 in., mortar joints	40	1.92
$\frac{1}{2}$ in., mastic-adhered	10	0.48
Concrete block:		
Lightweight, unreinforced, 4 in.	20	0.96
6 in.	25	1.20
8 in.	30	1.44
Heavy, reinforced, grouted, 6 in.	45	2.15
8 in.	60	2.87
12 in.	85	4.07

[a] Average weight per square foot of surface, except as noted. Values given as /in. or /mm must be multiplied by actual thickness of material.

Because of its permanent character, dead load affects design in the following ways:

- Designers always include dead load in loading combinations, except when investigating singular effects, such as deflections due to live load only.

- Over time, dead load causes sag (requiring reduction of design stresses in wood structures), develops long-term, continuing settlements in some soils, and produces creep effects in concrete structures.

- Dead load contributes some unique responses, such as the stabilizing effects that resist uplift and overturn due to wind forces.

Although designers can determine a material's weight, the complexity of most building construction means designers can computate only approximations of dead loads; as a result, design for structural behaviors is an approximate science. Do not use such complexity as an excuse for computational sloppiness, but you may temper your concern for precise computations.

Live Loads

Live load technically includes all nonpermanent loadings. However, the term usually refers only to the vertical gravity loadings on roof and floor surfaces. Although loads occur in combination with the dead loads, they are generally random in character, so designers must deal with them as potential contributors to various loading combinations.

Roof Load. Roofs are designed to support a uniformly distributed live load, which includes both snow accumulation and the general loads that accompany roof construction and maintenance. (Specified by local building codes, snow loads are based on local snowfalls.)

Table 12.2 gives the minimum roof live-load requirements specified by the 1994 edition of the UBC. Note the adjustments for roof slope and for total roof surface area supported by a structural element. The latter accounts for the decrease in likelihood of total surface loading as the size of the surface area increases.

Roof surfaces also must withstand wind pressure, as specified by building codes, which are based on local wind histories. For very light roof construction, wind's upward (suction) effect may exceed the dead load and result in a net upward lifting force.

The term *flat roof* is a misnomer; all roofs must slope, allowing some water drainage. The minimum required pitch is usually $\frac{1}{4}$ in./ft, or a slope of approximately 1:50. When designing roof surfaces that are close to flat, be sure to prevent *ponding*, where the

TABLE 12.2   Minimum Roof Live Loads[1]

ROOF SLOPE	METHOD 1			METHOD 2		
	Tributary Loaded Area in Square Feet for Any Structural Member			Uniform Load[2]	Rate of Reduction r (percentage)	Maximum Reduction R (percentage)
	0 to 200	201 to 600	Over 600			
	× 0.0929 for m² × 0.0479 for kN/m²					
1. Flat[3] or rise less than 4 units vertical in 12 units horizontal (33.3% slope). Arch or dome with rise less than one eighth of span	20	16	12	20	.08	40
2. Rise 4 units vertical to less than 12 units vertical in 12 units horizontal (33% to less than 100% slope). Arch or dome with rise one eighth of span to less than three eighths of span	16	14	12	16	.06	25
3. Rise 12 units vertical in 12 units horizontal (100% slope) and greater. Arch or dome with rise three eighths of span or greater	12	12	12	12		
4. Awnings except cloth covered[4]	5	5	5	5	No reductions permitted	
5. Greenhouses, lath houses and agricultural buildings[5]	10	10	10	10		

[1]Where snow loads occur, the roof structure shall be designed for such loads as determined by the building official. See Section 1605.4. For special-purpose roofs, see Section 1605.5.
[2]See Section 1606 for live load reductions. The rate of reduction r in Section 1606 Formula (6-1) shall be as indicated in the table. The maximum reduction R shall not exceed the value indicated in the table.
[3]A flat roof is any roof with a slope of less than $1/4$ unit vertical in 12 units horizontal (2% slope). The live load for flat roofs is in addition to the ponding load required by Section 1605.6.
[4]As defined in Section 3206.
[5]See Section 1605.5 for concentrated load requirements for greenhouse roof members.
Source: Adapted from the *Uniform Building Code,* 1994 ed. (Ref. 1), with permission of the publishers, the International Conference of Building Officials.

weight of water on the roof surface causes deflection of the supporting structure, which in turn allows more water accumulation (a pond), causing more deflection, and so on, resulting in an accelerated collapse condition.

Floor Load.   Floors must support all effects created by the occupancy; they must support, for example, the weights of occupants, furniture, equipment, stored materials, and so on. All building codes list minimum live loads for various occupancies, but different codes specify different live loads, so always use the local code. Table 12.3 contains floor loads given by the UBC.

TABLE 12.3 Minimum Floor Loads

USE OR OCCUPANCY		UNIFORM LOAD[1] (pounds)	CONCEN-TRATED LOAD (pounds)
Category	Description	× 0.004 48 for kN	
1. Access floor systems	Office use	50	2,000[2]
	Computer use	100	2,000[2]
2. Armories		150	0
3. Assembly areas[3] and auditoriums and balconies therewith	Fixed seating areas	50	0
	Movable seating and other areas	100	0
	Stage areas and enclosed platforms	125	0
4. Cornices and marquees		60[4]	0
5. Exit facilities[5]		100	0[6]
6. Garages	General storage and/or repair	100	7
	Private or pleasure-type motor vehicle storage	50	7
7. Hospitals	Wards and rooms	40	1,000[2]
8. Libraries	Reading rooms	60	1,000[2]
	Stack rooms	125	1,500[2]
9. Manufacturing	Light	75	2,000[2]
	Heavy	125	3,000[2]
10. Offices		50	2,000[2]
11. Printing plants	Press rooms	150	2,500[2]
	Composing and linotype rooms	100	2,000[2]
12. Residential[8]	Basic floor area	40	0[6]
	Exterior balconies	60[4]	0
	Decks	40[4]	0
13. Restrooms[9]			
14. Reviewing stands, grandstands, bleachers, and folding and telescoping seating		100	0
15. Roof decks	Same as area served or for the type of occupancy accommodated		
16. Schools	Classrooms	40	1,000[2]
17. Sidewalks and driveways	Public access	250	7
18. Storage	Light	125	
	Heavy	250	
19. Stores		100	3,000[2]
20. Pedestrian bridges and walkways		100	

[1]See Section 1606 for live load reductions.
[2]See Section 1604.3, first paragraph, for area of load application.
[3]Assembly areas include such occupancies as dance halls, drill rooms, gymnasiums, playgrounds, plazas, terraces and similar occupancies which are generally accessible to the public.
[4]When snow loads occur that are in excess of the design conditions, the structure shall be designed to support the loads due to the increased loads caused by drift buildup or a greater snow design as determined by the building official. See Section 1605.4. For special-purpose roofs, see Section 1605.5.
[5]Exit facilities shall include such uses as corridors serving an occupant load of 10 or more persons, exterior exit balconies, stairways, fire escapes and similar uses.
[6]Individual stair treads shall be designed to support a 300-pound (1.33 kN) concentrated load placed in a position which would cause maximum stress. Stair stringers may be designed for the uniform load set forth in the table.
[7]See Section 1604.3, second paragraph, for concentrated loads. See Table 16-B for vehicle barriers.
[8]Residential occupancies include private dwellings, apartments and hotel guest rooms.
[9]Restroom loads shall not be less than the load for the occupancy with which they are associated, but need not exceed 50 pounds per square foot (2.4 kN/m²).

Source: Adapted from the *Uniform Building Code*, 1994 ed. (Ref. 1), with permission of the publishers, the International Conference of Building Officials.

Although expressed as uniform loads, code-required values are usually large enough to account for ordinary concentrations. For offices, parking garages, and some other occupancies, however, codes often require designers to consider a specified concentrated load as well as the distributed loading. And designers must provide special detail when buildings are to contain heavy machinery, stored materials, or other contents of unusual weight.

When structural framing members support large areas, most codes allow designers to reduce the total live load. The UBC gives the following method for determining the reduction permitted for beams, trusses, or columns that support large floor areas:

Except for floors in places of assembly (e.g., theaters), and except for live loads greater than 100 psf [4.79 kN/m^2], the design live load on a member may be reduced in accordance with the formula

$$R = 0.08 (A - 150)$$

$$[R = 0.86 (A - 14)]$$

The reduction shall not exceed 40 percent for horizontal members or for vertical members receiving load from one level only, 60 percent for other vertical members, nor R as determined by

$$R = 23.1 \left(1 + \frac{D}{L}\right)$$

In these formulas

R = reduction (in percent)
A = area of floor support by a member
D = unit dead load/sq ft of supported area
L = unit live load/sq ft of supported area

In certain building types (e.g., office buildings), partitions are not permanently fixed. To provide for this flexibility, add 15 to 20 psf [0.72 to 0.96 kN/m^2] to other dead loads.

Lateral Loads

The term *lateral load* usually refers to the horizontal forces induced by wind and earthquakes on stationary structures. Design criteria and methods regarding lateral loads are refined continuously as designers learn more. Model building codes, such as the UBC, reflect currently recommended practices.

A complete discussion of lateral loads is beyond the scope of this book. Instead I summarize some of the design criteria specified in the latest UBC edition. I demonstrate how to apply such criteria later in this chapter, in the building structural design examples. For a more extensive discussion, refer to *Simplified Building Design for Wind and Earthquake Forces* (Ref. 9).

Wind.   Many codes contain simple criteria for wind design. In regions where wind is a major problem, however, local codes are usually more extensive. One of the most up-to-date and complex standards for wind design is in *Minimum Design Loads for Buildings and Other Structures* (Ref. 2).

When designing for wind effects on buildings, you must keep in mind a large number of architectural and structural concerns. I discuss some of these concerns in the following sections.

Basic Wind Speed.   The maximum wind speed (i.e., velocity); based on recorded wind histories and adjusted to match statistical likelihood. For the continental United States, designers use wind speeds listed in UBC, Figure No. 4. *Note:* Speeds are recorded at 10 m (approximately 33 ft) above ground level.

Wind Exposure.   The conditions of the terrain surrounding the building site. The UBC defines three categories (B, C, and D). Condition C refers to sites surrounded by at least one-half mile of flat, open terrain. Condition B means that buildings, forests, or ground-surface irregularities at least 20 ft tall cover at least 20 percent of the area 1 mile around the site. Condition D refers to sites by the sea and other special locations.

Note: The ASCE Standard describes four conditions (A, B, C, and D).

Wind Stagnation Pressure (q$_s$). The basic reference equivalent static pressure based on critical local wind speed. UBC Table No. 16-F is based on the following formula (given in the ASCE Standard):

$$q_s = 0.00256V^2$$

Example 1. For a wind speed of 100 mph,

$$q_s = 0.00256V^2 = 0.00256(100)^2$$
$$= 25.6 \text{ psf } [1.23 \text{ kPa}]$$

which is rounded off to 26 psf in the UBC table.

Design Wind Pressure (P). The equivalent static pressure to be applied normal to the building's exterior surfaces. Determined from the formula (UBC Section 1618)

$$P = C_e C_q q_s I_w$$

where P = design wind pressure (in psf)
C_e = combined height, exposure, and gust factor coefficient; given in UBC Table No. 16-G
C_e = pressure coefficient for the structure (or portion of structure under consideration); given in UBC Table No. 16-H
q_s = wind stagnation pressure at 30 ft; given in UBC Table No. 16-F
I_w = importance factor; is 1.15 for facilities essential for public health and safety (such as hospitals and government buildings) and buildings with hazardous contents, is 1.0 for all other buildings

The design wind pressure may be positive (inward) or negative (outward, suction) on a surface.

Design Methods. The UBC details two methods for applying design wind pressures when designing the primary bracing system:

> *Method 1 (Normal Force Method).* In this method wind pressures are assumed to act simultaneously normal to all exterior surfaces. This method is required for gabled rigid frames and may be used for any structure.
>
> *Method 2 (Projected Area Method).* In this method the total wind effect equals a single inward (positive) horizontal pressure acting on the projected building profile plus an outward (negative, upward) pressure acting on the full projected area of the building in plan. This method may be used for any structure shorter than 200 ft, except for gabled rigid frames. In the past, building codes used this method only.

For design of individual elements, refer to UBC Table No. 16-H for a C_q coefficient (used in determining P).

Uplift. May occur as a general effect, involving the entire roof or even the whole building, or as a local phenomenon (e.g., as generated by the overturning moment on a single shear wall). In general, both design methods account for uplift.

Overturning Moment. Most codes require that the ratio of the dead load resisting moment (called the *restoring moment, stabilizing moment*, etc.) to the overturning moment be 1.5 or greater. If the ratio is less than 1.5, a building must have sufficient anchorage to resist uplift effects.

Overturning is critical for

- Relatively tall and slender tower structures
- Individual bracing units in buildings braced by shear walls, trussed bents, and rigid-frame bents

Use Method 2 for this investigation, except for very tall buildings and gabled rigid frames.

Drift. The horizontal deflection due to lateral loads. Drift criteria is usually limited to the drift of a single story (in other words, horizontal movement of one level with respect to the story above or below). Although the UBC does not provide wind drift limits, other standards do. A common recommendation is to limit story drift to 0.005 times the story height (which is the UBC limit for seismic drift); for masonry structures drift is sometimes limited to 0.0025 times the story height. As for other situations involving structural deformations, designers must consider effects on the building construction—thus the detailing of curtain walls or interior partitions may affect drift limits.

Special Problems. The general design criteria given in most codes applies to ordinary buildings. More thorough investigation is recommended (and sometimes required) for special circumstances, including the following:

Tall Buildings. Designers must consider wind speeds and unusual wind phenomena at upper elevations.
Flexible Structures. Designers must investigate the effects of vibration or flutter. In addition, they must double-check whether such structures can withstand normal movements.
Unusual Shapes. Designers must study whether open structures, structures with large overhangs or other projections, and structures with complex shapes cause special wind effects. Some codes advise (even require) wind-tunnel testing.

Earthquakes. During an earthquake a building shakes. Because back-and-forth (horizontal) movements typically are more violent and destabilize buildings, structural design for earthquakes mostly accounts for lateral forces.

Lateral forces are generated by the building's weight—or, more specifically, by the building's mass, which represents both an inertial resistance to movement and the source for kinetic energy once the building is actually in motion. When using the equivalent static force method, designers consider the building structure to be loaded by a set of horizontal forces that consist of some fraction of the building weight. (Visualize the building being rotated vertically 90° to form a cantilever beam, with the ground as the fixed end and with a load equal to the building weight.)

In general, designing for the horizontal force effects of earthquakes is similar to designing for the horizontal force effects of wind; designers use the same basic types of lateral bracing (shear walls, trussed bents, rigid frames, etc.) to resist both. In the main a bracing system developed for wind bracing most likely will serve well for earthquake resistance.

Because the criteria and procedures are very complex, I do not illustrate the design for earthquake effects in the examples in this chapter. Nevertheless, the elements and systems developed for lateral bracing in these design examples are usually applicable to situations where earthquakes are a predominant concern.

For structural investigation, a major difference between seismic loads and wind loads is in their true dynamic effects: critical wind force usually is represented by a major, one-direction punch from a gust, while earthquakes represent rapid back-and-forth actions. However, once the dynamic effects are translated into equivalent static forces, design concerns for the respective bracing systems are very similar, involving considerations for shear, overturning, horizontal sliding, and so on.

For a detailed explanation of earthquake effects and illustrations of investigation by the equivalent static force method, refer to *Simplified Building Design for Wind and Earthquake Forces* (Ref. 9).

12.3 BUILDING ONE

Figure 12.1 shows a one-story, box-shaped building intended for commercial occupancy. As the detail section (Figure 12.1*d*) shows, the exterior walls are principally reinforced masonry with concrete blocks (i.e., concrete masonry units or CMUs), while the roof structure consists of light steel trusses supporting a formed sheet steel deck. Later in this section I discuss other construction options using different steel elements.

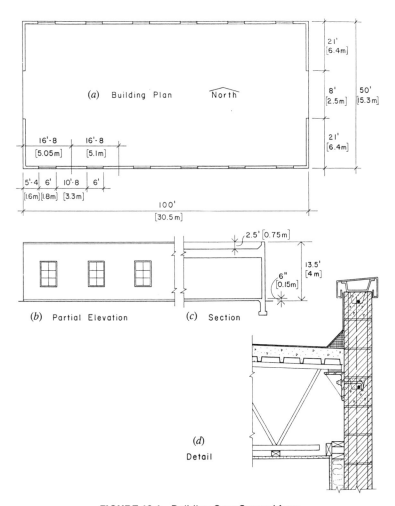

FIGURE 12.1 Building One: General form.

General Considerations

I assume the following data for design:

> Roof dead load = 15 psf, not including the weight of the structure
>
> Roof live load = 20 psf, reducible for large supported areas

Figure 12.1*d* indicates that the wall continues above the roof to create a parapet. Figure 12.1*d* also shows that the steel trusses are supported at the wall face. The joist span is approximately 48 ft.
    Figure 12.1*c* indicates that the building has a flat roof surface. Further, the roof deck sits directly on top of the trusses, and the ceiling is directly attached to the bottom of the trusses. *Note:* To ensure reasonable drainage, the roof surface must slope at least $\frac{1}{4}$ inch per foot (2 percent), but I assume a constant truss depth for design purposes.

The Roof Structure

Clear Spanning Roof. Spacing of the open-web joists must be coordinated with the roof deck and ceiling construction details. For a trial design, I assume a spacing of 4 ft. From Table 6.5, with deck units typically achieving three spans or more, I learn that the lightest deck in the table (22 gage) may be used. The deck configuration (i.e., rib width) depends on the materials placed on top of the deck and the means used to attach the deck to the supports.
    Adding the deck's weight to the other roof dead load produces a total dead load of 17 psf for the superimposed load on the joists. As I illustrated in Section 6.10, the design for a K-series joist is as follows:

Joist live load = 4(20) = 80 lb/ft (or plf)
Total load = 4(20 + 17) = 148 plf + the joist weight

From Table 6.2, the following choices are avilable for the 48 ft span.

- 24K9 at 12.0 plf, total load = 148 + 12 = 160 plf (less than the table value, 211 plf)
- 28K6 at 11.4 plf, total load = 148 + 11.4 = 159.4 plf (less than the table value of 184 plf)

Although the 28K6 is the lighter choice, you may want to use a deeper joist. For example, if the ceiling is directly attached to the bottoms of the joists, a deeper joist provides more space for building service elements. Another reason: Deflection is reduced if you use a deeper joist. Although pushing the live load deflection to the limit—(1/360)(48 × 12) = 1.9 in.—may not be critical for the roof surface, doing so can present problems for the structure's underside, leading to ceiling sag or difficulties with nonstructural walls built up to the ceiling. In sum, a 30K7 at 12.3 plf results in considerably less deflection without adding much weight.

Note: Table 6.2 is an abridged table; many more joist sizes exist. In this book I mean only to demonstrate how to use such references.

Specifications for open-web joists give requirements for end-support details and lateral bracing (see Ref. 7). If you use the 30K7 for the 48 ft span, for example, you also must use four rows of bridging.

In Figure 12.1*d* the support indicated for the joists results in an eccentric load on the wall, which induces bending in the wall. Figure 12.2 shows a common alternative detail for the roof-to-wall joint: the joists sit directly on the wall, and the joist top chord extends to form a short cantilever.

Roof with Interior Columns. If this building does not require a clear spanning roof structure, it may be possible to use some interior columns and create a framing system with quite modest spans.

Figure 12.3*a* shows a framing plan for a system that uses columns at 16 ft 8 in. on center in each direction. You can use either short-span joists or a longer span deck, as indicated on the plan. This span exceeds the capability of the deck with 1.5 in. ribs, but decks with deeper ribs are available.

Figure 12.3*b* shows a second possible framing arrangement: the deck spans the other direction and only two rows of beams are used. This arrangement allows for wider column spacing, which increases the beam spans but eliminates 60 percent of the interior columns and their footings (a major cost savings).

Sometimes you can make beams in continuous rows simulate a continuous beam action without moment-resistive connections. For example, you can use beam splice joints off the columns (see Figure 12.3*c*); doing so allows for relatively simple connections but reduces deflections.

For the beam in Figure 12.3*b*, but assuming a slightly heavier deck, an approximate dead load of 20 psf results in a beam load of

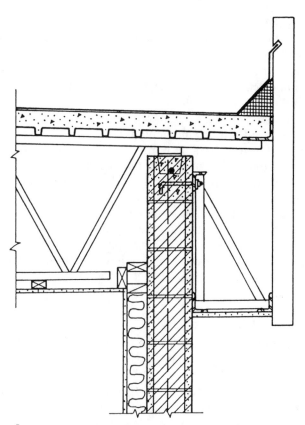

FIGURE 12.2   Building One: Variation of roof-to-wall detail. For comparison, see Figure 12.2*d*.

$$w = 16.67(20 + 16) = 600 \text{ plf} + \text{ the beam} \simeq 640 \text{ plf}$$

(Note that the beam periphery of $33.3 \times 16.67 = 555 \text{ ft}^2$ qualifies the beam for a live-load reduction.)

The simple-beam bending moment for the 33.3 ft span is

$$M = \frac{w L^2}{8} = \frac{(0.640)(33.3)^2}{8} = 88.9 \text{ kip-ft}$$

From Table B.1, the lightest W shape beam permitted is a W 16 × 31. From Figure 6.3, the total load deflection is approximately

L/240, which is usually not critical for roof structures. Furthermore, the live-load deflection is less than one-half this amount.

If you build the three-span beam with three simple-spanning segments, the detail at the top of the column is as in Figure 12.4, but the framing detail in Figure 12.3, with the beam-to-beam connection off the column, is better.

The investigation in Figure 12.5, which shows the beam reactions, shear, and bending moments, is made with a beam splice 4 ft from the column. The center portion of the three-span beam thus becomes a simple span of 25.33 ft. The end reactions for the center portion become loads on the ends of the extensions of the outer portions.

The maximum bending moment is 71.09 kip-ft, or approximately 80 percent of that for the simple-span beam. Table B.1 indicates that you can reduce the shape to a W 16 × 26.

The support reaction of 22.458 kips is the total design load for the column. Assuming a height of 10 ft and a K of 1, you can choose among the following column choices:

- W 4 × 13 (see Table 7.2)
- 3 in. pipe (nominal, standard weight) (see Table 7.3)
- 3 in. square tube, $\frac{3}{16}$ in. thick (see Table 7.4)

Gabled Roof (with Trusses)   Figure 12.6 is another possible Building One roof structure: a gabled (double-sloped) roof form. The building profile in Figure 12.6a includes a series of trusses spaced at plan intervals (as shown for the beam-and-column rows in Figure 12.3a).

Investigation of the Trusses.   Figure 12.6b shows the truss form, and Figure 12.7 lists the results of an algebraic analysis for a unit loading. Of course, the true unit loading for the truss is derived from the construction form.

Note: This design is based only on gravity loading because the usual allowable stress increases for wind loads lessen the effects of wind load, except in regions hit by very strong windstorms. Still, if a deck is directly supported by the trusses, the top chords may experience more bending.

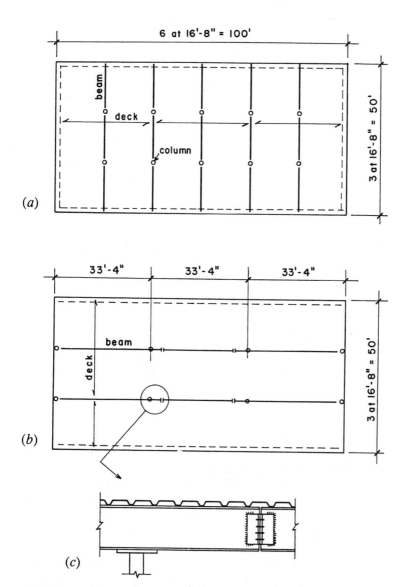

FIGURE 12.3 Building One: Options for roof framing with interior columns.

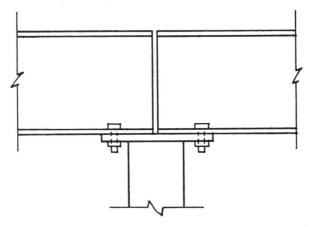

FIGURE 12.4   Framing detail at the top of the steel column with simple beam action. For comparison, see Figure 12.3*c*.

Design of the Steel Truss.   Figure 12.6*d* shows the use of double-angle members with joints developed with gusset plates. The top chord extends to form a cantilever. (For clarity sake, the detail shows only the major structural elements. Additional construction would be required to develop the roofing, ceiling, and soffit.)

In trusses of this size, designers commonly extend the chords without joints for as long as possible. Available lengths depend on members sizes and the usual lengths stocked by local fabricators. Figure 12.6*c* shows a possible layout comprising a two-piece top chord and a two-piece bottom chord. The longer top chord is 36 ft, plus the overhang—a size that may be difficult to obtain if the angles are small.

The roof construction illustrated in Figure 12.6*d* has a long-span steel deck that bears directly on the top chord. Using such a deck simplifies the framing by eliminating the need for intermediate framing between trusses. For the truss spacing 16 ft 8 in. (see Figure 12.3*a*), the deck is quite light and this system is feasible. However, the direct bearing of the deck adds a spanning function to the top chord, the chords will be considerably heavier.

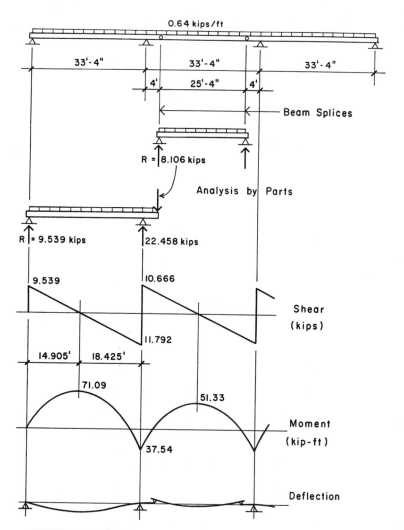

FIGURE 12.5 Development of the continuous beam with internal pins.

Given a roof live load of 20 psf and a total roof dead load of 25 psf (deck + insulation + roofing substrate + tile roofing), the unit load on the top chord is

$$w = (20 + 25)(16.67) = 750 \text{ lb/ft}$$

To be conservative, assume a simple beam moment:

$$M = \frac{wL^2}{8} = \frac{(0.750)(10)^2}{8} = 9.375 \text{ kip-ft}$$

Using an allowable bending stress of 22 ksi,

$$\text{Required } S = \frac{M}{F_b} = \frac{(9.375)(12)}{22} = 5.11 \text{ in.}^2$$

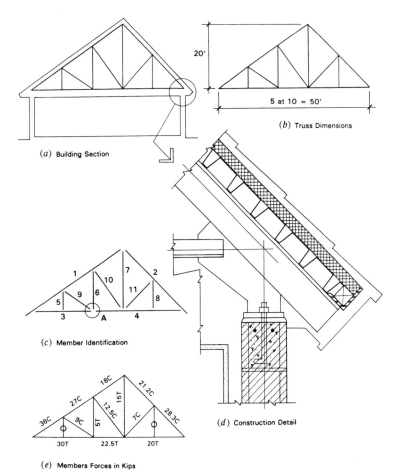

(a) Building Section

(b) Truss Dimensions

20'

5 at 10 = 50'

(c) Member Identification

(d) Construction Detail

(e) Members Forces in Kips

FIGURE 12.6 Building One: Form of the gabled truss roof structure.

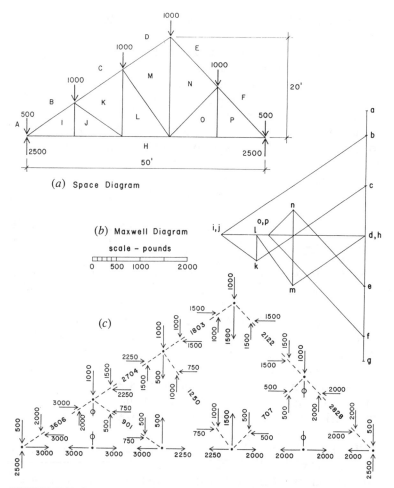

FIGURE 12.7   Investigation of the truss for gravity load. The space diagram (a) indicates the truss layout and the unit loading. The Maxwell diagram (b) is a graphical analysis for the internal forces in the truss members with the unit loading. The separated joint diagram (c) shows the individual concentric force systems at the joints, the components of forces in the sloping members, and the net force in all members. *Note:* I use this truss for the example analyses in Section 10.4.

If the bending task is about three-fourths of the effort for the chord, you can determine an approximate size by looking for a pair of double-angles with a section modulus of at least 7.5. Table A.5

presents a possibility: a pair of $6 \times 4 \times \frac{1}{2}$ in. angles, with a section modulus of 8.67 in.[2]

Before speculating any more, investigate for the internal forces in the truss members. Figure 12.7 shows concentrated forces of 1000 lb each at the top chord joints. (*Note:* Designers typically assume this form of loading even though the actual load is distributed along the top chord (roof load) and the bottom chord (ceiling load).) If the total of the live load, roof dead load, ceiling dead load, and truss weight is approximately 60 psf, the single joint load is

$$P = (60)(10)(16.67) = 10,000 \text{ lb.}$$

This load is ten times the load shown in Figure 12.7, so the internal forces for the gravity loading are ten times those shown in Figure 12.7 (refer to Figure 12.6e). Table 12.4 summarizes the design of all the truss members, but the top chords; tension members reflect a desire for welded joints and a minimum angle leg $\frac{3}{8}$ in. thick, while compression members come from AISC tables.

When designing light trusses, you can derive a minimum-size member from the layout, dimensions, magnitude of forces, or joint details. For example, designers may use angle legs just wide enough to accommodate bolts or thick enough to accommodate fillet welds. Minimum L/r ratios for members are another source of minimum criteria. Minimum design sometimes results in a poor truss design, with the combination of form, size, and proposed construction type not meshing well.

The truss chords in question can consist of structural tees with most gusset plates eliminated, except possibly at the supports. If you can transport the trusses to the site and erect them as one piece, you can weld all the joints within the trusses in the shop. Otherwise, you must develop a scheme for dividing the truss and splicing the separate pieces in the field; most likely you must use high-strength bolts for the field connections.

To derive an approximate design of the top chord, consider the following combined function equation:

$$\frac{f_a}{F_a} + \frac{f_b}{F_b} = 1$$

TABLE 12.4 Design of the Truss Members

Truss Member			
No.	Force (kips)	Length (ft)	Member Choice (All Double-Angles)
1	36C	12	Combined bending and compression member $6 \times 4 \times \frac{1}{2}$
2	28.3C	14.2	Max. $L/r = 200$, min. $r = 0.85$ $6 \times 4 \times \frac{1}{2}$
3	30T	10	Max. $L/r = 240$, min. $r = 0.5$ $3 \times 2\frac{1}{2} \times \frac{3}{8}$
4	22.5T	10	$3 \times 2\frac{1}{2} \times \frac{3}{8}$
5	0	6.67	$2\frac{1}{2} \times 2\frac{1}{2} \times \frac{3}{8}$
6	5T	13.33	Max. $L/r = 240$, min. $r = 0.6$ $2\frac{1}{2} \times 2\frac{1}{2} \times \frac{3}{8}$
7	15T	20	Max. $L/r = 240$, min. $r = 1.0$ $3\frac{1}{2} \times 3\frac{1}{2} \times \frac{3}{8}$
8	0	10	$2\frac{1}{2} \times 2\frac{1}{2} \times \frac{3}{8}$
9	9C	12	$2\frac{1}{2} \times 2\frac{1}{2} \times \frac{3}{8}$
10	12.5C	16.67	Max. $L/r = 200$, min. $r = 1.0$ $3\frac{1}{2} \times 2\frac{1}{2} \times \frac{3}{8}$
11	7C	14.2	Min. $r = 0.85$ $3 \times 2\frac{1}{2} \times \frac{3}{8}$

You can calculate this simple form of the straight-line interaction graph, by considering two ratios: the value of the compression required to that of the compression capacity, and the actual bending requirement (required S) to the actual S. Thus

Required compression = 36.06 kips
Capacity (load table in AISC Manual) = 137 kips
    Ratio = 36.06/137 = 0.263
Required S = 5.11 in.^{3}
Actual S (property table in Ref. 3) = 8.67 in.^{3}
    Ratio = 5.11/8.67 = 0.589
Sum of the ratios = 0.852

These computations indicate that the member choices are reasonable, although AISC specifications required a more elaborate investigation.

Figure 12.8*a* shows a possible configuration for Joint A. Since the bottom chord splice occurs at this point (see Figure 12.6*c*), the chord is shown as discontinuous at the joint. If the truss is site-assembled in two parts, this joint may accommodate field connecting; Figure 12.8*b* shows an alternative, with the splice joint achieved with bolts.

Development of the Roof Structure.   Developing a complete roof construction entails designing a roof deck system—and probably a framed ceiling system as well. You can directly support both by the truss chords, or you can develop an intermediate set of framing members supported at the truss joints (Figure 12.9 shows framing plans for both options).

When trusses are closely spaced (as with open-web joists), the system in Figure 12.9*a* is more common. When the spacing is too wide, you cannot use a clear-spanning deck. Thanks to the additional framing, however, you eliminate some bending in the truss chords.

Fire codes permitting, these trusses may be exposed to view on the interior—that is, with no suspended ceiling below the trusses. However, structural supports often are required for lighting, fire sprinkler piping, HVAC ducts and fixtures, and so on. To support

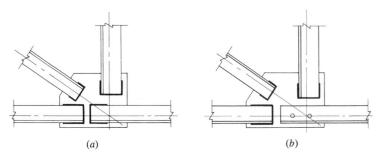

(*a*)   (*b*)

FIGURE 12.8   Form of a typical truss joint using double-angle members with gusset plates and welded connections. (*a*) Indicates a fully welded joint. (*b*) Shows a possibility for a field splice joint, with steel bolts used for the field-connected member.

these elements, designers can choose among many options, including direct support by truss members or wire hangers inserted through the roof deck.

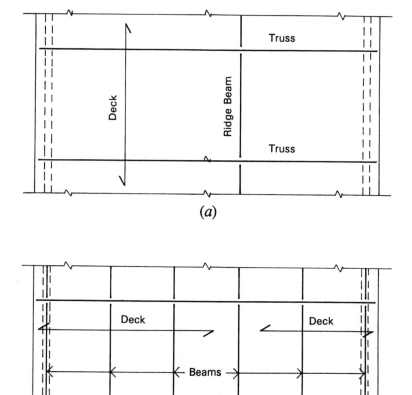

FIGURE 12.9 Partial framing plan for the truss roof structure. (a) With a structural deck spanning between trusses and no secondary framing. (b) With beams between trusses, supported at the truss joints to avoid bending in the truss chords.

Gabled Roof (with Welded Steel Bents) Figure 12.10 illustrates a gable-form roof with a welded bent structure. The bent is fabricated from steel plates, welded together to form I shaped cross sections for the members (see Figures 12.10c and d). This roof is quite common, whether built from steel, glue-laminated wood, or precast concrete.

When this structure is constructed as shown, with pin-type connections at the ridge of the roof and at the column base, the reaction forces are statically determinate, and so its design is relatively simple. Moreover, because many steel product companies produce this very basic element—in fact, it is a standard catalog item—you can quickly develop a complete design by choosing among standard commercial options. In any event, I do not show the complete investigation and design for this structure; if desired, you can find examples in many texts on steel or timber design.

Note: Remember to develop some horizontal resistance to outward movement at the base of the columns—a need common to arches and gable-form structures. The simplest means for doing so is to use a tie across the building to the opposite column, if spatial form and construction details permit its installation.

Lateral Forces

To enable this size and form of building to resist wind or earthquake forces, designers typically use a *box system*, consisting of shear walls and a horizontal diaphragm roof. If planning of solid walls does not provide sufficient potential for development of shear walls, you may use a trussed frame (i.e., braced frame) or rigid frame.

Designing steel decks as horizontal diaphragms is routine, done with data supplied by deck manufacturers—individual deck products and connecting elements have rated capacities. Elements of the steel frame are used as diaphragm chords, collectors, ties, and drag struts. For example, Building One's masonry walls could be used as shear walls, while the lightest steel deck would suffice for diaphragm action if adequately attached to the framing. In fact, the minimum construction permitted by codes and industry standards is adequate for anything other than major windstorm conditions.

Note: The building examples that follow present greater lateral load challenges.

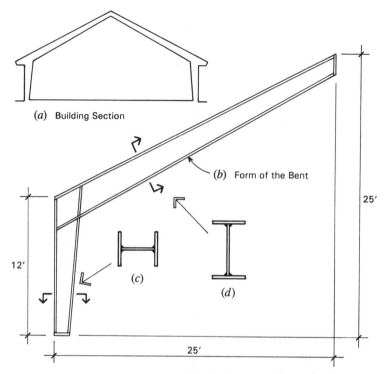

(a) Building Section

(b) Form of the Bent

25′

12′

(c)

(d)

25′

FIGURE 12.10 Building One: Form of steel rigid-frame bents.

12.4 BUILDING TWO

Note: In this section I use the same live-load and dead-load data given for Building One.

Figure 12.11 shows a partial framing plan for the roof structure of a one-story industrial building. The system, with 48-ft-square bays, is repeated a number of times in each direction. The plan indicates a series of girders supported by columns, a series of joists perpendicular to the girders, and a roof deck supported by the joists. The spacing of the joists, a critical planning decision for this system, affects the following:

Load on a Single Joist. The larger the spacing, the more load a joist must resist.

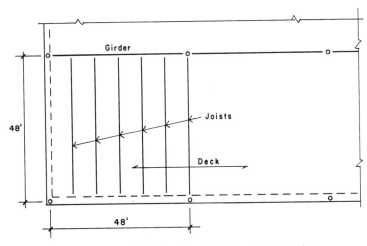

FIGURE 12.11 Building Two: Partial roof framing plan.

The Deck. Plywood panels have short spans but steel decks depend on the rib depth. Proprietary products have specific limits, so check the manufacturer's data.

Point Loads on Girders. If the girders are trusses, loads must be placed at upper chord joints.

When selecting the deck, joists, girders, and columns for this structure, designers get to choose from a wide range of possible combinations. In this section I outline two common solutions.

Alternative One: Joists and W Shape Girders

This system uses joists placed at 4 ft centers and supported by W shape girders. *Note:* For this span and loading, the joist choices are the same as for Building One.

Because a single interior girder supports a large area, the roof live load drops to 12 psf (see Table 12.2). If you assume a total dead load with the joists of 20 psf, the total load for the girder is 32 psf and the linear load on the girder (considered a uniformly distributed load) is

$$w = (32)(48) = 1540 \text{ plf, or } 1.54 \text{ kips/ft}$$

In fact, the joist loads constitute point loads at 4 ft on the girder, but the difference in the maximum bending moment is quite small. If you add some assumed weight for the girder, a reasonable approximation is 1.6 kips/ft. For simple beam action, the maximum bending moment is

$$M = \frac{wL^2}{8} = \frac{(1.6)(48)^2}{8} = 518.4 \text{ kip-ft}$$

From Table B.1, the lightest possible shape is a W 30 × 99.

You can use the scheme with the beam joints off the column, as with Building One. If you can similarly reduce the maximum moment 20 percent, the moment drops to (0.80)(518.4) = 415 kip-ft. Now the lightest shape possible is a W 27 × 84, boosting the unit loading to 1.624 plf and the design moment to approximately 421 kip-ft.

For the column, the total load is (1.624)(48) = 77.95 kips. Given a clear height of 20 ft and a *K* of 1, the following are possible:

- W 8 × 31 (see Table 7.2)
- 8 in. standard pipe (see Table 7.3)

Note: The AISC Manual lists larger tubes, up to 16 in. square. A 6-in-square tube with wall thickness of $\frac{5}{16}$ in. will work for this example.

Alternative Two: Joists and Joist Girders

This system uses trusses in place of W shape girders.

Open-web joist manufacturers design and market the *joist girder* (refer to Section 6.10) for this type system. Your "design" work, as a result, is limited to determining layout, loading, and girder depth.

The spaced point loads of the joists determine a horizontal module for the girder. This dimension should match the truss depth so that the truss panels do not approach the extremes shown in Figure 12.12. A rule of thumb: The depth of the truss in inches should approximate the span in feet (in other words, span/depth = 12). For this example, a depth of 48 in. produces square truss panels (not necessarily ideal, but within the reasonable range of proportions).

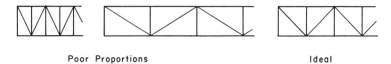

Poor Proportions   Ideal

FIGURE 12.12   Layouts of truss form for the girders.

With 4 ft joist spacing, the truss has 12 panels, and the unit load at the truss-girder panel point on the top chord is equal to the total load on one joist. Given a total average dead load of 20 psf for the joist, but a live load reduced to 12 psf for the girder design (see Table 12.2), this panel point load is

$$4(12 + 20) \times 48 = 6144 \text{ lb, or } 6.144 \text{ kips}$$

As I described in Section 6.10, the joist girder designation is thus 48G12N6.144K, indicating a 48-in.-deep girder with 12 spaces for supported joists (actually 11 joists) and a joist load of 6.144 kips.

Column options are generally the same as those for Alternative One. Connecting a W shape girder to a column is different than connecting a truss, however, so the same column shape (cross section) may affect the two systems differently.

Note: This basic system is an alternate design for Building Three (see Figure 12.31).

12.5   BUILDING THREE

Building Three is a modest-size office building generally known as a low rise (see Figure 12.13). For such buildings designers can choose among many construction forms. Nonetheless, in any place at a given time, a few forms are most popular.

General Considerations

This type building usually requires some modular planning to coordinate columns, window mullions, and interior partitions. Such modular coordination also may include ceiling construction, lighting, HVAC elements, and the systems enabling access to elec-

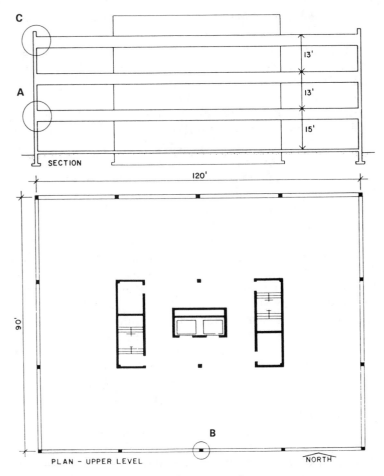

FIGURE 12.13 Building Three: General form.

tric power, phones, and other signal wiring. No single magic dimension exists; various designers use and advocate different dimensions between 3 and 5 ft. To establish a reference dimension, select a curtain wall system, interior modular partitioning, or an integrated ceiling system.

Usually, a building built as an investment property whose occupancy may vary over the building's life, must easily accommodate future redevelopment of the building interior. As a result, designers develop a basic construction design with as few permanent structural elements as possible—columns, floors, roof, exterior walls, and interior walls that enclose stairs, elevators, restrooms, and service risers. Everything else should be nonstructural or demountable in nature, if possible.

Moreover, designers try to space columns as wide as possible to reduce the number of freestanding columns. In fact, they strive for a column-free interior if the distance from a central core (i.e., grouped permanent elements) to the outside walls is not too great for a single span. Columns at the building perimeter are not a problem, so designers sometimes add columns there to improve the building's resistance to gravity loads and lateral loads.

Complicating matters, the space between the underside of a suspended ceiling and the top of a floor or roof structure typically must contain not only basic construction elements, but also elements of the structural, HVAC, electrical, communication, lighting, and fire-fighting systems. Before you can integrate all such elements, you must correctly assess how much total space is required. Unfortunately, once you establish the depth permitted for the spanning structure and the general level-to-level vertical building height, these dimensions are not easy to change later if the detailed design of an enclosed system indicates a need for more space.

Although providing ample space for such building elements eases the work of subsystem designers, taller exterior walls, stairs, elevators, and service risers cost more.

Designers also must choose a basic construction form for the exterior walls. For the column-framed structure, you must integrate the columns and the nonstructural infill wall. One basic construction form incorporates the columns into the wall, with windows developed in horizontal strips between the columns (see Figures 12.14 and 12.15). Because the exterior column and spandrel covers develop a general continuous surface, the window units appear as "punched" holes in the wall.

The windows in this example are not parts of a continuous curtain wall system; instead they are essentially individual units, placed in and supported by the general wall system. The curtain wall is developed as a stud-and-surfacing system, not unlike the typical light wood stud wall system. The studs are light-gage steel,

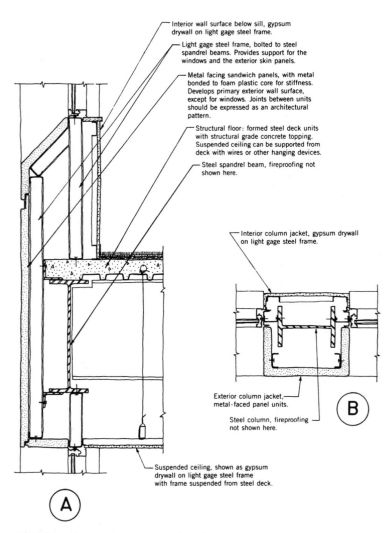

Interior wall surface below sill, gypsum drywall on light gage steel frame.

Light gage steel frame, bolted to steel spandrel beams. Provides support for the windows and the exterior skin panels.

Metal facing sandwich panels, with metal bonded to foam plastic core for stiffness. Develops primary exterior wall surface, except for windows. Joints between units should be expressed as an architectural pattern.

Structural floor: formed steel deck units with structural grade concrete topping. Suspended ceiling can be supported from deck with wires or other hanging devices.

Steel spandrel beam, fireproofing not shown here.

Interior column jacket, gypsum drywall on light gage steel frame.

Exterior column jacket, metal-faced panel units.

Steel column, fireproofing not shown here.

Suspended ceiling, shown as gypsum drywall on light gage steel frame with frame suspended from steel deck.

FIGURE 12.14   Building Three: Wall, floor, and column construction at the typical office floor.

349

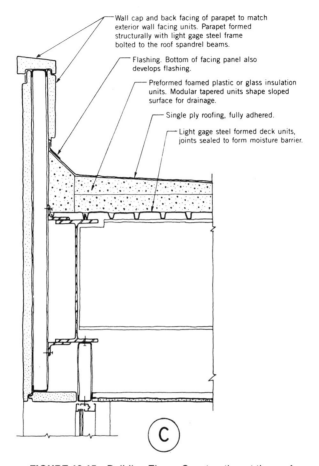

Wall cap and back facing of parapet to match exterior wall facing units. Parapet formed structurally with light gage steel frame bolted to the roof spandrel beams.

Flashing. Bottom of facing panel also develops flashing.

Preformed foamed plastic or glass insulation units. Modular tapered units shape sloped surface for drainage.

Single ply roofing, fully adhered.

Light gage steel formed deck units, joints sealed to form moisture barrier.

C

FIGURE 12.15 Building Three: Construction at the roof.

the exterior covering is a system of metal-faced sandwich panel units, and the interior covering (where required) is gypsum drywall, attached to the metal studs with screws.

Detail A shows a considerable interstitial void space that easily can contain service elements (e.g., the electrical system) as well as the usual insulation. In cold climates, this space often holds a perimeter hot water heating system.

Structural Alternatives

Designers can choose among many structural options, including all steel frame, concrete frame, and masonry bearing wall systems. You even can choose the light wood frame if the total floor area and zoning requirements permit its use. Which structural elements you choose depend mostly on the desired plan form, type of window arrangements, and clear spans required for the building interior.

At the height of Building Three, the basic structure is often steel or reinforced concrete. In this section I describe structures that use steel columns with different combinations of horizontal-spanning steel structures and lateral bracing systems.

When designing the structural system, take into account both gravity and lateral force systems. For gravity, designers must develop horizontal-spanning systems for the roof and upper floors and stack vertical supporting elements. For lateral loads, designers choose among the following common lateral bracing systems (see Figure 12.16):

Core Shear Walls (Figure 12.16a). Using solid walls around core elements (stairs, elevators, restrooms, duct shafts) produces a very rigid vertical structure; the rest of the construction may lean on this rigid core.

Truss-Braced Core. Similar to the shear wall core, except trussed bents replace solid walls.

Perimeter Shear Walls (Figure 12.16b). Turns the building into a tubelike structure. Walls may be either structurally continuous and pierced by holes for windows and doors or built as individual, linked piers between vertical strips of openings.

Mixed Exterior and Interior Shear Walls or Trussed Bents. A mixture of walls and trussed bents; used when the perimeter core systems are not feasible.

Full Rigid-Frame Bents (Figure 12.16c). Uses all the available bents described by vertical planes of columns and beams.

Perimeter Rigid-Frame Bents (Figure 12.16d). Uses only the columns and spandrel beams in the exterior wall planes, resulting in only two bents in each direction.

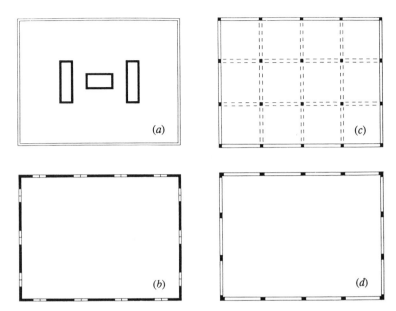

FIGURE 12.16 Building Three: Options for vertical elements of the lateral bracing system. (a) Braced core; shear walls or trussing. (b) Braced perimeter; shear walls or trussing. (c) Fully developed, three-dimensional rigid frame. (d) Perimeter rigid frames.

Each system has some advantages and disadvantages, structurally and architecturally.

In the following sections I detail schemes that use a truss-braced core and a full rigid-frame bent. Both use steel W shape columns for vertical support. I also present two schemes for the horizontal roof and floor structures: a beam-and-girder system using all W shapes and a truss system with open-web joists and joist girders.

Design Criteria

I use the following data:

Building Code: 1994 UBC (Ref. 1)
Live Loads:
Roof: Table 12.1

Floor: from Table 12.3, 50 psf minimum for office areas, 100 psf for lobbies and corridors, 20 psf for movable partitions
Wind: map speed of 80 mph, exposure B
Assumed construction loads:
Floor finish: 5 psf
Ceilings, lights, ducts: 15 psf
Walls (average surface weight):
Interior, permanent: 15 psf
Exterior curtain wall: 25 psf
Steel for rolled shapes: ASTM A36, F_y = 36 ksi

The Beam-and-Girder Floor Structure

Figure 12.17 shows a typical framing system for the upper floor. Note how the spacing between rolled steel beams relates to the column spacing: The beams are 7.5 ft on center, so the beams not on column lines are supported by column-line girders. Thus girders support three-fourths of the beams and columns directly supported the remainder. The beams in turn support a one-way spanning deck.

This basic system has several variables:

- *Beam Spacing.* Affects the deck span and beam loading.
- *Deck.* A variety available, as discussed later.
- *Beam/Column Relationship.* Permits possible development of vertical bents in both directions.
- *Column Orientation.* The W shape has a strong axis and accommodates framing differently in different directions.
- *Fire Protection.* See Section 3.5.

Figure 12.17 shows not only the common elements but also several special beams required at the building core. However, I limit the discussions that follow to the common elements—that is, the members labeled beam and girder in Figure 12.17.

In a speculative rental building, each floor may have a different plan. Because designers cannot predict the locations of offices and corridors—each requires different live loads—many design for

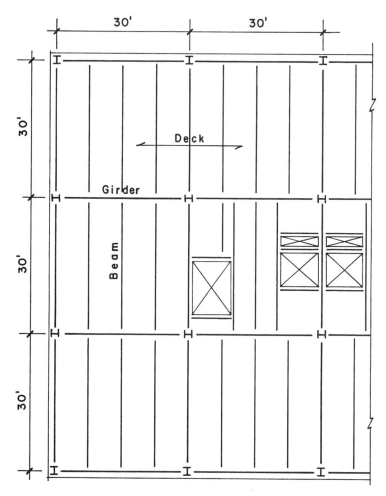

FIGURE 12.17   Building Three: Partial framing plan for the steel floor structure for the upper levels.

general loading combinations. For the design work in this section, I use the following data:

For the deck: live load = 100 psf
For the beams: live load = 80 psf, with 20 psf added to dead load for movable partitions

For girders and columns: live load = 50 psf, with 20 psf added to dead load

The Structural Deck. Several floor deck options are possible. Designers must consider not only structural concerns, including gravity and lateral loads, but also

- How to protect the steel from fire
- How to accommodate wiring, piping, and ducts
- How to attach finish floor, roofing, and ceiling constructions

Office buildings often must house electrical power and communication networks built into the wall and floor constructions. If the structural floor deck, for example, is a concrete slab (sitecast or precast), you can bury such power and communication networks in the nonstructural fill placed on top of the structural slab. If you use a steel deck, you can hide some wiring in closed cells of the formed sheet steel deck units.

For this example, the selected deck is a steel deck with 1.5-in.-deep ribs; on top of the deck is cast a lightweight concrete fill with a minimum depth of 2.5 in. over the steel units. The unit average dead weight of this deck depends on the thickness of the sheet steel, the profile of the deck folds, and the unit density of the concrete fill. For this example, I assume that the deck's total dead load is 50 psf.

Although industry standards exist (see Ref. 8), I urge you to obtain data from deck manufacturers. Industry standards are guidelines, but actual manufactured products vary.

The Common Beam. The beam in Figure 12.17 spans 30 ft. It also carries a 7.5-ft-wide load strip, which allows the following live-load reduction (see Section 12.2):

$$R = 0.08(A - 150) = 0.08(225 - 150) = 6\%$$

Thus the beam loading is

Live load = 7.5(0.94)(80) = 564 lb/lineal ft (or plf)
Dead load = 7.5(50) = 525 plf + beam weight $\simeq$ 560 plf

Total unit load = 1124 plf or 1.124 kips/ft

Then

$$M = \frac{wL^2}{8} = \frac{1.124(30)^2}{8} = 126.45 \text{ kip-ft}$$

Allowing for the beam weight, Table B.1 yields the following possible choices: W 16 × 45, W 18 × 46, or W 21 × 44. The deeper shape obviously produces the least deflection, although in this case the live-load deflection for the 16 in. shape is within the usual limit (see Figure 6.3). Meanwhile, the deck should provide virtually continuous lateral bracing of the top (compression) flange.

This beam is the typical member; you design other beams—including the column-line beams, the spandrels, and so on—as necessary.

The Common Girder.   Figure 12.8 shows the loading condition for the girder—but as generated only by the supported beams. *Note:* You can ignore the effect of the girder's weight (a uniformly distributed load) because the girder weight is a minor load.

The girder carries three beams and thus has a total load periphery of 3(225) = 675 ft². Allowable live-load reduction is

$$R = 0.08(675 - 150) = 42\%$$

However, because the maximum reduction for horizontal-spanning structures is 40 percent, the unit beam load is thus

Dead load = 0.570(30) = 17.1 kips
Live load = 0.60(0.050)(7.5)(30) = 6.75 kips
Total load = 17.1 + 6.75 = 23.85 kips ≃ 24 kips

To select an appropriate member, you must heed various data. After determining the maximum moment (see Figure 12.18), consult Table B.1 or Figure B.1 to determine acceptable choices. Given that this member is laterally braced at 7.5 ft intervals, the lightest choices are W 24 × 84 and W 27 × 84. You also can use W 21 × 93 or W 30 × 90. The deeper members have less deflection, but the shallower ones allow more room for building service elements in the floor/ceiling enclosed space.

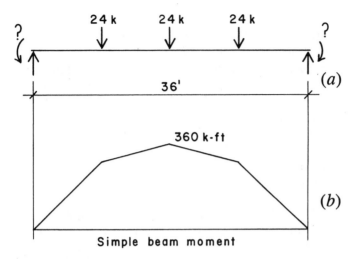

FIGURE 12.18 Loading condition for gravity loads on the girder (a) from the supported beams. End restraints or continuity reduces the maximum bending moment from that of the simple beam moment (b).

Although you can compute deflections with formulas that recognize the true form of loading, you can find approximate deflection values by using an equivalent load derived from the maximum moment, as I discussed in Section 6.6. For this example, the equivalent uniform load (EUL) is

$$M = \frac{WL}{8} = 360 \text{ kip-ft}$$

$$W = \frac{8M}{L} = \frac{8(360)}{30} = 96 \text{ kips}$$

Use this hypothetical uniformly distributed load with simpler formulas to find an approximate deflection.

Although you should investigate deflection of individual elements, other deflection issues are important, including the following:

Floor Bounce. Bounce, which involves the stiffness and the fundamental period of spanning elements, relates to the deck, the beams, or both. In general, using the static deflection limits ensures a reasonable lack of bounce, but just about anything that increases stiffness improves the situation.

Load Transfer to Nonstructural Walls. With the building construction completed, live-load deflections may cause spanning members to bear on nonstructural construction. Reducing deflections may help, but designers often must design special details to attach the structure to the nonstructural construction.

Deflection During Construction. Girder deflection plus beam deflection creates a cumulative deflection at the center of a column bay (see Figure 12.19). Not only may this deflection be critical for live load, but it also can create problems during construction. If you install the steel beams and steel deck dead flat, then later construction—for example, concrete fill—causes deflection from the flat condition. One solution is to camber (i.e., bow upward) the beams in the shop so that they deflect to flat.

Column Design for Gravity Loads

When designing steel columns, you must consider both gravity and lateral loads.

Gravity loads are based on a column's *periphery*, usually defined as the area of supported surface on each level supported. Loads are actually delivered to columns by beams and girders, but designers use the peripheral area to tabulate loads and determine live-load reductions.

If beams are attached rigidly to columns with moment-resistive connections—as with rigid-frame bents—then gravity loads also cause bending moments and shears in the columns. Otherwise, gravity loads are essentially only axial compressive loads.

How much you involve columns to resist lateral loads depends on lateral bracing system. If you use trussed bents, some columns function as chords in the vertically cantilevered trussed bents, thus adding some compressive forces and possibly causing some reversals with net tension in the columns. If columns are parts of rigid-frame bents, the same chord actions are present but the columns also are subject to bending moments and shears from the rigid-frame lateral actions.

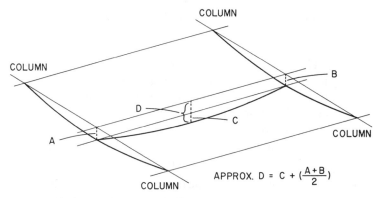

FIGURE 12.19 Cumulative deflection of the floor structure: A and B are girder deflections; C is the beam deflection; D is the true total deflection at the midpoint between columns.

Whatever the lateral force actions may do, the columns alone must resist gravity load effects. In this section I ignore lateral loads. Doing so leads to reference column designs that you can modify (but not reduce) when designing the lateral resistive system. (Later in this section I discuss two lateral bracing systems: a trussed bent system and a rigid-frame system.)

Column loading depends on framing arrangements and column locations. To completely design all columns, you must tabulate the loading for each case. In this section, however, I focus on three cases: a corner column, an intermediate exterior column, and a hypothetical interior column. (*Note:* For the interior column, I assume a general periphery of 900 sq ft—general roof or floor area. Of course, the floor plan in Figure 12.13 shows that all interior columns fall within the core area, so no such column shall exist. However, I show later how all interior columns affect lateral force systems, so this tabulation is worthwhile, also yielding a column size appropriate when designing for lateral forces.)

Figure 12.20 is a common form of table designers use to determine column loads. For the exterior columns, the table lists three separate load determinations, corresponding to three-story-high columns. For the interior columns, the table assumes that a fourth story—a rooftop structure (e.g., penthouse)—sits above the core.

Tables like the one in Figure 12.20 help designers determine the following:

- Dead load on the periphery at each level, which is calculated by multiplying the area by an assumed average dead load per square foot. Designers may use the loads determined while designing the horizontal structure.
- Live load on the periphery areas.
- Live-load reduction for each story; based on the total supported periphery areas above that story.
- Other directly supported dead loads, such as column weight and any permanent walls within the load periphery.

Level	Load Source	Corner Column 225 ft²			Intermediate Exterior Column 450 ft²			Interior Column 900 ft²		
		DL	LL	Total	DL	LL	Total	DL	LL	Total
P'hse.	Roof							8	5	
Roof	Wall							5		
	Total/level							13	5	
	Design load									18
Roof	Roof	9	5		18	9		36	18	
	Wall	10			10			10		
	Column	3			3			3		
	Total/level	22	5		31	9		49	23	
	Design load			27			40			72
3rd	Floor	16	11		32	23		63	45	
Floor	Wall	10			10			10		
	Column	3			3			3		
	Total/level	51	16		76	32		125	68	
	LL reduction	24%	12		60%	13		60%	27	
	Design load			63			89			152
2nd	Floor	16	11		32	23		63	45	
Floor	Wall	11			11			11		
	Column	4			4			4		
	Total/level	82	27		123	55		203	113	
	LL reduction	42%	16		60%	22		60%	45	
	Design load			98			145			248

FIGURE 12.20 Building Three: Table of column loads. Assumes a rooftop penthouse at the building core. The interior column is hypothetical, ignoring actual core conditions at the upper floors.

- Total load at each level.
- Design load per story; based on the total accumulation from all levels supported.

Figure 12.20 reflects the following assumptions:

Roof unit live load = 20 psf (reducible)
Roof dead load = 40 psf (estimated, based on the similar floor construction)
Penthouse floor live load = 100 psf (for equipment, average)
Penthouse floor dead load = 50 psf
Floor live load = 50 psf (reducible)
Floor dead load = 70 psf (includes partitions)
Interior walls weigh 15 psf/ft^2 of wall surface
Exterior walls weigh average of 25 psf/ft^2 of wall surface

Figure 12.21 summarizes the designs for the three columns. For the pin-connected frame, a K factor of 1.0 is assumed and the full

Level	Story	Unbraced Height (ft)	Corner Column		Intermediate Exterior Column		Interior Column	
			Design Load (kips)	Column Choices	Design Load (kips)	Column Choices	Design Load (kips)	Column Choices
Roof								
	3rd	13	27	W10X33	40	W10X33	72	W10X39
3rd Floor								
	2nd	13	63	W10X33	89	W10X33	152	W10X39
2nd Floor					Assumed location of column splice			
	1st	15	98	W10X33	145	W10X33	248	W10X49
1st Floor								

FIGURE 12.21 Building Three: Summary of column designs based on Figure 12.20.

story heights are used as the unbraced column lengths.

Although column loads are low in the upper stories, I suggest that the W shapes be at least 10 in. for the following reasons:

Form of the Horizontal Framing Members, and Type of Connections between Columns and Horizontal Framing. H shaped columns usually must facilitate framing in both directions, with beams connected both to column flanges and webs. With standard framing connections for field bolting to the columns (see Section 11.4), a minimum beam depth and flange width are required to enable practical installation of connecting angles and bolts. Figure 12.22, a plan view at the corner column (note the beams framing from both directions into the W 10 $\times$ 33 column), shows limits for the angle sizes, bolt sizes, and beam flange widths (unless flanges are cut back).

Splices in the Multistory Column. If the building is too tall for a single-piece column, you must use a splice somewhere. Stacking column pieces is much easier when the two pieces are similar in size (see Figure 7.10).

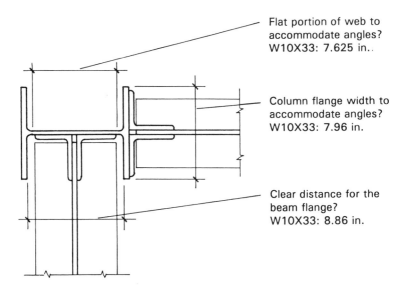

Flat portion of web to accommodate angles?
W10X33: 7.625 in.

Column flange width to accommodate angles?
W10X33: 7.96 in.

Clear distance for the beam flange?
W10X33: 8.86 in.

FIGURE 12.22 Dimensional considerations for accommodation of beam framing in both directions at the W shape columns.

Steel Size. Long steel pieces are tough to transport; once on site they are not easy to handle. The smaller the member's cross section, the shorter the piece that is feasible to handle.

The W 10 × 33 is often considered the minimum column; it is the lightest shape with an 8-in.-wide flange (see Figure 12.22). In Figure 12.21 a splice goes 3 ft above the second-floor level (a convenient, waist-high height for the erection crew); the two column pieces are approximately 18 and 23 ft long, lengths readily available and easy to handle.

When I discussed the design of a column base plate in Section 7.13, I used the shape selection and design load shown in Figure 12.21 for the first-story interior column.

Design of Lateral Bracing Systems

The Trussed Bent   In Figure 12.23, a partial framing plan for the core area, some columns are positioned off the 30 ft grid. These columns exist to help define the vertical bents needed for the

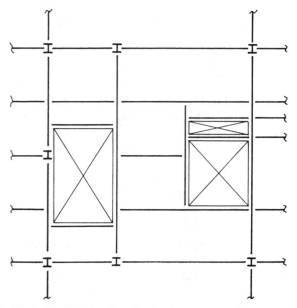

FIGURE 12.23   Modified framing plan for development of the trussed bents at the core.

trussed bracing system shown in Figure 12.24. With slender diagonal members, the X-bracing behaves as I described in Section 8.4—the tension diagonals function alone. In total, four vertical, cantilevered, determinate trusses brace the building in each direction.

Given the symmetrical building exterior form and the symmetrically placed core bracing, these trusses (in conjunction with the horizontal roof and upper-floor structures) should effectively resist horizontal forces due to wind. In the work that follows, I illustrate the design process, using criteria for wind loading from the 1994 UBC (Ref. 1).

For the total wind force, the code permits designers to use the profile method, which defines the pressure on a vertical surface as

$$p = C_e C_q q_s I$$

Assuming wind speed of 80 mph and exposure condition B, and excluding any special concern for the I factor, Table 12.5 lists the wind pressures at various Building Three height zones. To investigate the lateral bracing system, translate the wind pressures on the exterior wall surface into edge loadings for the roof and upper-floor diaphragms (see Figure 12.25). The vertical structure spans to distribute the external wind pressure.

The accumulated forces noted as H_1, H_2, and H_3 in Figure 12.25 are shown applied to a vertical trussed bent in Figure 12.26a; for an east-west bent, loads are determined by multiplying the diaphragm edge loading by the building width and then dividing by the number of bents. Thus

$$H_1 = (185.7)(92)/4 = 4271 \text{ lb}$$
$$H_2 = (215.5)(92)/4 = 4957 \text{ lb}$$
$$H_3 = (193.6)(92)/4 = 4453 \text{ lb}$$

Figure 12.26b shows the truss loading together with the reaction forces at the supports. Figure 12.26c shows the internal forces in the truss members resulting from this loading, with force values in pounds and sense indicated by C for compression and T for tension.

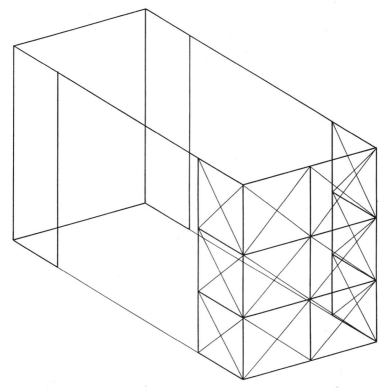

FIGURE 12.24 General form of the trussed bent bracing system at the core.

TABLE 12.5 Design Wind Pressure for Building Three (Exposure Condition B)[a]

Building Surface Zone	Height Above Ground (ft)	C_e	C_q	Pressure p (psf)
1	0–15	0.62	1.3	13.2
2	15–20	0.67	1.3	14.3
3	20–25	0.72	1.3	15.4
4	25–30	0.76	1.3	16.2
5	30–40	0.84	1.3	17.9
6	40–60	0.95	1.4	21.8

[a]Horizontally directed pressure on vertical surface: $p = C_e \times C_q \times 16.4$ psf.

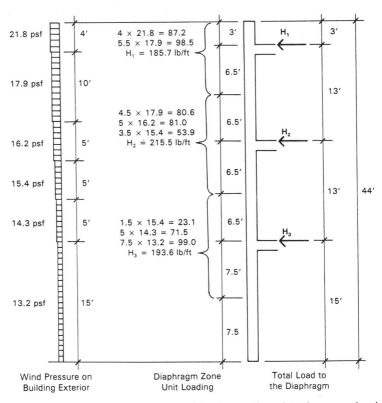

| Wind Pressure on Building Exterior | Diaphragm Zone Unit Loading | Total Load to the Diaphragm |

FIGURE 12.25 Building Three: Wind loads transferred to the upper-level horizontal diaphragms (roof and floor decks). See Table 12.5 for the design wind pressures on the building exterior. Diaphragm zones are defined by column midpoints. To find a total load on the diaphragm at a level, multiply the diaphragm zone unit load per foot by the width of the building.

You can use the forces in the diagonals to design tension members, allowing the usual increase of stress for the ASD method. Be sure to add the compression forces in the columns to the gravity loads to see whether this load combination is critical. Then compare the uplift tension force at the windward column with the dead load to see whether you must redesign the column base for a tension anchorage force.

You should add the horizontal forces to the beams in the core framing and investigate for the combined bending and compres-

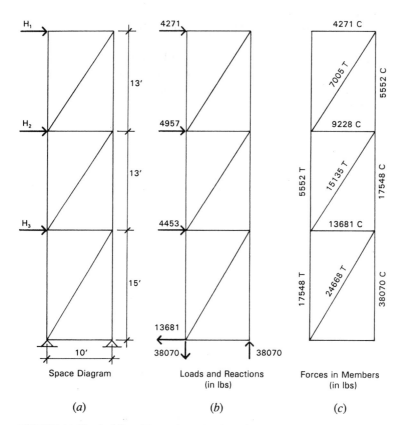

FIGURE 12.26 Building Three: Investigation of an east-west core bent. (a) Layout and loading of the bent; loads shown are one-fourth of the total diaphragm loads. (b) External forces (loads and reactions) on the cantilevered truss. (c) Internal forces in the truss members.

sion. Since beams often are weak on their minor axes (y-axis), you may need to add some framing members at right angles to brace the beams against lateral buckling.

When designing diagonals and their connections to the beam and column frame, you must consider their form and the wall construction in which they are embedded. (Figure 12.27 shows some possible details). If you use double-angles for the diagonals (a common truss form), the splice joint in Figure 12.27 is necessary where the two diagonals cross. If you use either single-angles or channel shapes for the diagonals, the members pass each other, back to back, at the center. *Note:* If loads are high, however, do not use single angles or channel shapes, which involve some degree of eccentricity in the members and connections and a single shear load on the bolts.

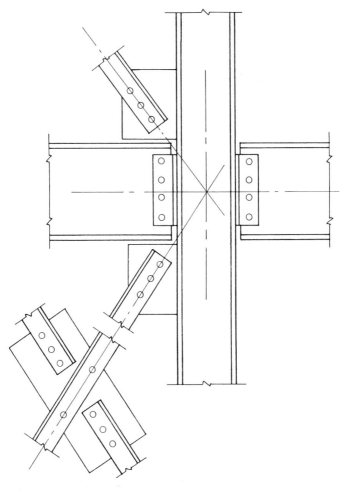

FIGURE 12.27 Details of the bent construction with bolted joints.

The following summarizes design considerations for a bottom diagonal that carries a wind load of 27.2 kips, assuming bolted connections with $\frac{3}{4}$ in. A325 bolts and a double-angle member.

Member length:

$$L = \sqrt{(10)^2 + (15)^2} = 18 \text{ ft}$$

For the tension member, a recommended minimum slenderness is represented by an L/r ratio of 300. Thus

$$\text{Minimum } r = \frac{18 \times 12}{300} = 0.72 \text{ in.}$$

At an allowable stress of 1.33(22) = 29.3 ksi on the gross section, the gross area of the pair of angles must be

$$A_g = \frac{24.7}{29.3} = 0.84 \text{ in.}^2$$

Assuming F_u = 58 ksi, and assuming an allowable stress on the net section of $0.50F_u$, the required net area at the bolt holes is

$$A_n = \frac{24.7}{1.33 \times 0.50 \times 58} = 0.64 \text{ in.}^2$$

Assuming a minimum angle leg for the $\frac{3}{4}$ in. bolts, and assuming a design hole size of $\frac{7}{8}$ in., the net width of the connected angle leg is

$$W = 2.5 - 0.875 = 1.625 \text{ in.}$$

If only the connected legs function in tension, the required thickness of the connected angle legs is

$$t = \frac{0.64}{2 \times 1.625} = 0.197 \text{ in.}$$

When rounded to $\frac{1}{4}$ in., this thickness is probably the minimum practical for use with this size bolt.

Table A.5 yields $2\frac{1}{2} \times 2 \times \frac{1}{4}$ in. angles with $\frac{3}{8}$ in. back-to-back spacing and a minimum r value for the x-x axis of 0.784 in.

Table 11.1 yields a value of 15 kips for one bolt in double shear, so only two bolts are required.

Alternative Floor Construction with Trusses

Figure 12.28, a framing plan for the upper floor of Building Three, indicates the use of open-web steel joists and joist girders. (*Note:* Although you can extend this construction to the core and the exterior spandrels, you also can use rolled shapes there.) Such construction allows various lateral bracing possibilities, including the trussed bent. Actually, this system is more suited to longer spans and lighter loads (see Building Two).

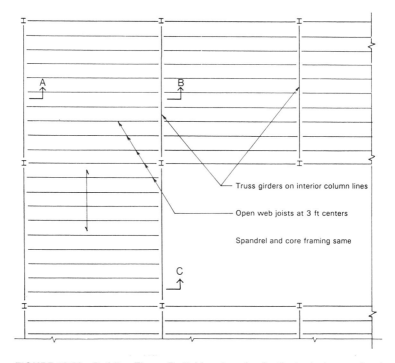

FIGURE 12.28 Building Three: Partial framing plan for the typical upper-level floor using open-web joists and joist girders. *Note:* Core framing is essentially the same as in Figure 12.17. Lateral bracing (bridging) for the joists is not shown.

One advantage of using the all-truss framing: Building service elements may pass freely within the enclosed space between the ceiling and the supported structure above. A disadvantage: The height of trusses usually demands greater structural depth, adding to building height—a problem that increases with the number of stories.

Design of the Open-Web Joists. The joist design summarized in Figure 12.29 uses the usual method of seeking the most efficient (in other words, lightest) member for selection. However, for floor construction, bounce is a major concern—especially with very light structures. For this reason, I recommend using the deepest feasible joists for floors. In general, increasing the depth (overall height of trusses) reduces both static deflection (sag) and dynamic deflection (bounce).

Note: I presented general concerns regarding open-web joists in Section 6.11.

Design of the Joist Girders. Often joists and girders are supplied and erected by a single contractor. Do not rely on industry standards (e.g., Ref. 7) but consult the specific manufacturer for data regarding design and construction details.

The pattern of joist girder members is somewhat fixed, relating to the spacing of supported joists. To ensure a reasonable proportion for the truss panel units, girder depth should be approximately equal to joist spacing.

In Figure 12.30, the assumed girder depth is 3 ft—a *minimum* depth for this span. Any additional depth reduces the amount of steel and improves deflection responses. However, for floor construction, this dimension is hard to bargain for.

Note: I discussed joist girders in Section 6.11.

Construction Details for the Truss Structure. Figure 12.31 shows some construction details of the trussed system. The deck shown is essentially the same as that for the scheme with W shape framing, although the shorter span may allow you to use a lighter sheet steel deck. However, the deck also affects diaphragm action, limiting the reduction possible.

Not only is the structure's height a problem, but the detail at the joist support requires the joists to sit atop the supporting members

Computations for the Open Web Steel Joists

Joists at 3' centers, span of 30' (+ or -)

Loads:

DL = (70 psf X 3') = 210 lb/ft (not including weight of joist)

LL = (100 psf X 3') = 300 lb/ft (not reduced)
This is a high load for the offices, but allows a corridor anywhere. Also helps reduce deflection to eliminate floor bounciness.

Total load = 210 + 300 = 510 lb/ft + joist weight

Selection of Joist from Table B.16:

Note that the table value for total load allowable includes the weight of the joist, which must be subtracted to obtain the load the joist can carry in addition to its weight.

Table data: total load = 510 lb/ft + joist, LL = 300 lb/ft, span = 30'

Options from Table B.16:

24K9, total allowable load = 544 - 12 = 532 lb/ft, OK

26K9, stronger than 24K9, only 0.2 lb/ft heavier

28K8, stronger, only 0.7 lb/ft heavier

30K7, stronger, only 0.3 lb/ft heavier

Note that the loading results in a shear limit for the joists, rather than a bending stress limit.

All of the joists listed are economically equivalent, so choice would probably be made with regard to dimensional considerations for the total depth of the floor construction. A shallower joist means a shorter story and less overall building height. A deeper joist gives more space in the floor construction for accommodation of ducts, etc. and probably the least floor bounce.

FIGURE 12.29 Summary of design for the open-web joist.

(in the all-W-shape system, the tops of beams and girders are level).

Thanks to the closely spaced open-web joists, the joists' bottom chords may directly support the ceiling construction. However, it is also possible to suspend the ceiling from the deck, as is generally required for an all-W-shape structure with widely spaced beams.

For the joist girder:

Use 40% LL reduction with LL of 50 psf.
Then, LL = 0.60(50) = 30 psf, and the total joist load is

(30 psf)(3 ft c/c)(30 ft span) = 2700 lb, or 2.7 kips

For DL, add partitions of 20 psf to other DL of 40 psf

(60 psf)(3 ft c/c)(30 ft) = 5400 lb

+ joist weight at (10 lb/ft)(30 ft) = 300 lb

Total DL = 5400 + 300 = 5700 lb, or 5.7 kips

Total load of one joist on the grider:

DL + LL = 2.7 + 5.7 = 8.4 kips

Girder specification for choice from manufacturer's load tables:
(See Ref. 11)

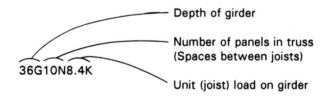

— Depth of girder

— Number of panels in truss
(Spaces between joists)

36G10N8.4K

— Unit (joist) load on girder

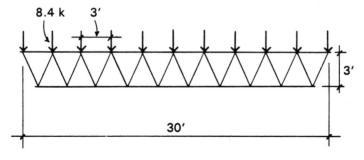

FIGURE 12.30 Summary of design for the joist girder.

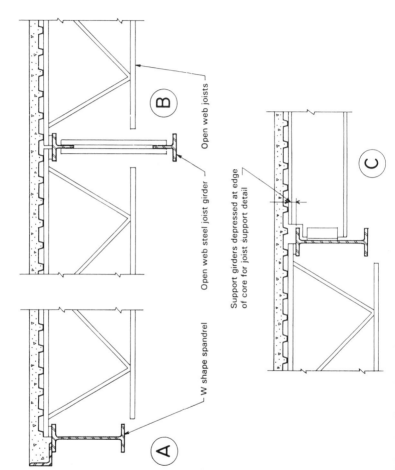

FIGURE 12.31 Details for the floor system with open-web joists and joist girders. For location of details, see the framing plan in Figure 12.28.

W shape spandrel

Open web steel joist girder

Open web joists

Support girders depressed at edge of core for joist support detail

A

B

C

Another issue: You usually must use a fire-resistive ceiling construction because it is not feasible to encase joists or girders in fireproofing material.

Rigid Frames. A critical concern for a multistory, multiple-bay rigid frame is its colulmns' lateral strength and stiffness. Because a building must be able to resist lateral forces in all directions, in many cases designers must consider how columns resist shear and bending in *two* directions (north-south and east-west, for example). W shape columns, for example, have considerably greater resistance on their major (x-x) axis versus their minor (y-y) axis, so how you orient W shape columns sometimes is a major structural planning decision.

Figure 12.32a shows a possible Building Three column orientation: two major bracing bents resist east-west and five shorter, less stiff bents resist north-south. The two stiff bents may well equal the resistance of the five shorter bents, giving the building a reasonably symmetrical response.

Figure 12.32b is a columnar plan designed to produce approximately symmetrical bents on the building perimeter. Figure 12.33 shows the form of such perimeter bracing.

With perimeter bracing, you can use deeper (and thus stiffer) spandrel beams—the restriction on depth that applies for interior beams does not exist at the exterior wall plane. You also can place more columns at the exterior (see Figure 12.32c) without compromising the building interior space. In fact, with deeper spandrels and closely spaced exterior columns, a very stiff perimeter bent is possible. Such a bent has so little flexing in the members that its behavior approaches that of a pierced wall, rather than a flexible frame.

At the expense of requiring much stronger (and heavier and/or larger) columns, plus moment-resistive connections, rigid-frame bracing offers significant architectural planning advantages, eliminating solid shear walls or truss diagonals in the walls. However, you must limit the frames' lateral deflection (i.e., drift) to prevent damage to nonstructural construction.

Note: I discussed rigid frames in Section 4.4 and Chapter 8.

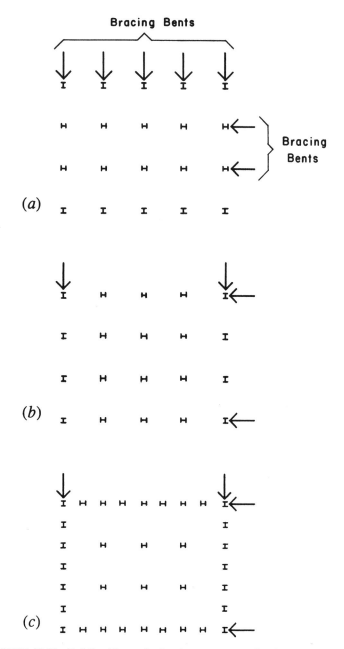

FIGURE 12.32 Building Three: Optional arrangements for the steel W shape columns for development of rigid-frame bents.

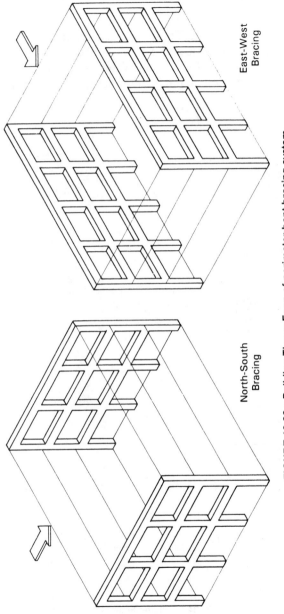

East-West
Bracing

North-South
Bracing

FIGURE 12.33 Building Three: Form of perimeter bent bracing system.

12.6 BUILDING FOUR

Figure 12.34 is Building Three adapted for mill construction. (Developed in the seventeenth century, mill construction consists of heavy masonry walls with an interior framed structure—originally made of timber, but later models used iron or steel—for

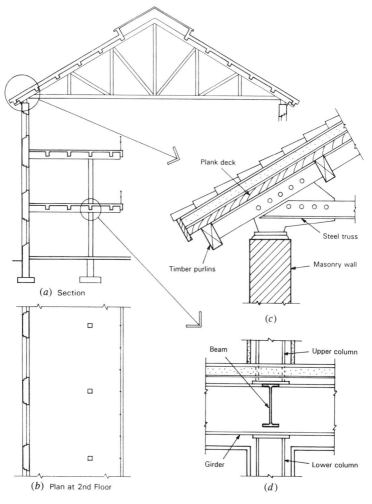

(a) Section

(b) Plan at 2nd Floor

(c)

(d)

FIGURE 12.34 Building Four: General form.

the floors and roofs.) The interior includes a high open space in the center, extending from the ground floor to the clear-spanning roof trusses. Upper floors are donut-shaped; a row of interior columns lines each side. An open corridor is formed on a cantilevered balcony that rings the open space.

Steel Floor and Columns

Figure 12.35 is a partial framing plan for Building Four's upper floor. Interior columns placed at 16 ft centers—matching the location of structural columns in the exterior masonry construction—support cantilevered steel girders that, in turn, support the 8-ft-wide balcony. Perpendicular are steel beams that support a steel deck with concrete fill.

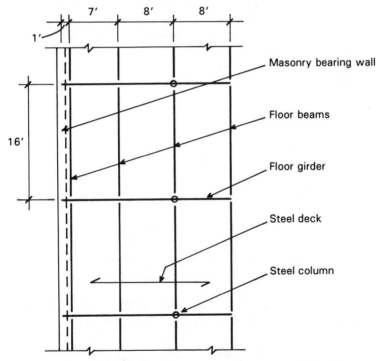

FIGURE 12.35 Building Four: Partial framing plan for upper-level floors.

As with Building Three, designing the steel beam and deck system for at least a 100 psf live load for the entire floor permits a future rearrangement and use of space. Thus, the only permanent structural elements are the steel columns and those elements necessary to enclose stairs, elevators, restrooms, and vertical service risers. As I determined for Building Three, I assume the total dead load for the floor to be 60 psf (includes the weights of all members except steel girders and columns).

You can choose the same deck used for Building Three—a sheet steel unit with 1.5-in.-high ribs and a 2.5 in. concrete fill over the deck units. When securely attached to each other, the units form a continuous deck; when welded to the top of the steel beams, the units help develop necessary diaphragm action for lateral loads. Interior beams carry a full 8-ft-wide strip of the deck, while beams at the exterior wall and at the balcony edge carry a one-half wide strip load from the deck. Because the beam at the inside face of the exterior wall prevents the wall from carrying the deck directly, the only load transfer to the exterior wall is through the girder. W shapes are used for the interior beams, and C (or channel) shapes are used for the edge beams.

For the interior beam, the strip load is

$$w = 8(100 + 60) = 1280 \text{ plf, or } 1.28 \text{ kip/ft}$$

and the maximum bending moment is

$$M = \frac{wL^2}{8} = \frac{(1.28)(16)^2}{8} = 40.96 \text{ kip-ft}$$

Table B.1 indicates that you can use a rather modest shape for this rather short span—choices include M 14 × 18, W 12 × 19, and W 10 × 22. The deeper beams are lighter but have very thin webs, so be sure to consider the details for end connections to the girders (see Figures 11.9e and f).

Using 22 ksi as a limit for bending stress (for unsymmetrical shapes), the required section modulus for the channels is

$$S = \frac{M}{F_b} = \frac{40.96 \times 12}{22} = 22.34 \text{ in.}^3$$

The AISC Manual indicates that you can use a C 12 × 25, with an *S* value of 24.1 in.[3] *Note:* Although it carries less load, this member is actually heavier than the W shapes for the interior beam, proving the structural efficiency of the W shapes.

When choosing beams, designers must coordinate their preferences with the floor and ceiling construction to ensure, among other things, that the steel structure is protected from fire.

Figure 12.36 summarizes the girder and column design. Note that the girder's critical bending moment is at the cantilever, which is also the most critical for static and dynamic (bounce) deflections. The 16-in.-deep shape may be adequate for static deflection and bending moment, but any extra depth will further stiffen the cantilever, reducing the balcony's bounce. (If the lunch-hour crowd wants to use the balcony as a running track, you probably should use at least a 27-in.-deep member.)

The cantilevered girders at the multistory columns pose a problem at the second floor. Ordinarily steel columns are made continuous, with beams framing into their sides, but Building Four's girders are continuous and the columns are interrupted (see Figure 12.34*d*). As a result, designers must pay extra attention to the vertical compression in the girder web (see Section 6.8).

To prevent bending in the girder flanges and vertical buckling of the girder web, add a thick plate into the open U-shaped area defined by the girder flanges and web on each side of the girder (see Figure 12.34*d*). Then fit individual story-high column segments with bearing plates at their tops and bottoms. Bolting the plates to the girders holds the frame in place during construction.

Steel Roof Truss

The roof structure uses 60-ft-span trusses, keeping the top floor interior free of columns. The detail in Figure 12.34*c* indicates the use of a steel truss with top chords formed from structural tees (actually, split W shapes) and interior members formed with double-angles. Joints are formed by welding the angle ends directly to the webs of the tee chords. Such trusses require special joints to ease erection at the supports, as well as at any interior field splice locations if the trusses cannot be shipped in one piece to the site.

FLOOR DECK

Assume steel deck with concrete fill. Use total floor dead load of 50 psf (same as for Building 3).

Use LL = 100 psf, office load + partitions = 75 psf, corridor = 100 psf. Permits location of corridor anywhere on floor. Select deck from manufacturer's catalogs.

BEAM, 8 ft c/c, 16 ft span

Total load/ft = 8(150) = 1200 lb/ft + beam, say 1230 lb/ft or 1.23 k/ft.

M = wL²/8 = (1.23)(16)²/8 = 39.36 k-ft.
Required S = M/F_b = (39.36 × 12)/24 = 19.68 in.³.

From Table A.1: W 12 × 26, S = 33.4 (lightest in table).
From Table B.1: W 12 × 19, S = 21.3.

Figure 6.12, deflection of 12 in. beam not critical on 16 ft span.

Beams at wall and at edge of cantilever carry less load, but may use same shapes (lightest in class in AISC).

GIRDER, loads and span as shown

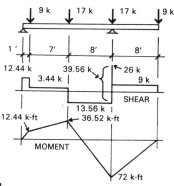

Area supported = 24 × 16 = 384 sq ft.
Use 80% Of LL, Total beam loads as shown.

Max. M = 72 k-ft, Min. req'd S = 36 in.³.

Table A.1: W 14 × 30 or W 16 × 36.
Table B.1: W 16 × 26, but L_c and L_u both less than 8 ft. W 16 × 31 probably OK. (Note: AISC shows both W 14 × 30 and W 16 × 31 OK. Use 16 in. beam to reduce bounce of cantilever.)

COLUMNS

2nd Story, 11'-8" high to bottom of girder, say 12 ft. Load = approx. 40 kips.

Pipe: Table 7.3, 4 in. nominal diameter.
Tube: Table 7.4, 4 × 4 × 3/16.
W Shape: Table 7.2, W 5 × 16, W 6 × 15, W 8 × 24.

1st Story, 13'-8" high to bottom of girder, say 14 ft. Load = approx. 80 kips.

Pipe: 6 in. nominal diameter.
Tube: 5 × 5 × 5/16.
W Shape: W 6 × 25 or W 8 × 24.

FIGURE 12.36 Building Four: Summary of design for the steel structure.

Figure 12.37 shows gravity load values for the top chord joints with a unit loading; multiply the values found for internal forces in the truss members with this loading by the true unit loading (as determined from the final construction details).

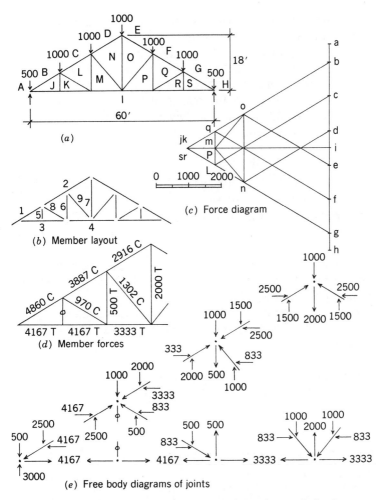

FIGURE 12.37 Investigation of the roof truss for gravity load.

The roof surface is generated by purlins that span between trusses and support a steel deck. If the purlins deliver their end reaction loads to the truss' top chord joints, you need not use heavy top chords. Although such a design complicates matters a bit, creating a connections traffic jam at this location, experienced steel detailers can handle the traffic. Placing the purlins atop the chords simplifies the intersection, but introduces a rollover stability problem for the tilted purlins. Figure 12.38 shows wind load values, based on UBC criteria. The load form is based on the roof slope and a minimum horizontal wind pressure of 20 psf at the roof level. Table 12.6 summarizes the results of the investigations illustrated in Figures 12.37 and 12.38. Given a sloped roof surface, the critical load combinations are

- Dead load plus live load.
- Dead load plus wind load when the forces in the member have the same sign (tension or compression).
- Dead load plus wind load when the net reverses the sign of the internal force produced by gravity load alone.

A complete analysis shows that wind load is not a critical concern for this structure, so you can design the truss members and connections for gravity loads alone.

Table 12.7 summarizes the design of the truss members for the loads in Table 12.6. When selecting members, designers must consider the following:

- Minimum thickness for interior angle members, based on the use of fillet welds for the truss joints (see Section 11.6).

TABLE 12.6   Design Forces for the Truss (in Pounds)

Member (see Figure 12.37)	Unit Gravity Load	Dead Load (3.6 × Unit)	Live Load (1.8 × Unit)	Wind Load	DL + LL
1	4860 C	17496 C	8748 C	3790 T	26244 C
2	3887 C	13994 C	6997 C	2450 T	20990 C
3	4167 T	15000 T	7500 T	1960 T/5600 C	22500 T
4	3333 T	12000 T	6000 T	3170 C	18000 T
5	(Zero force, all loadings)				
6	500 T	1800 T	900 T	820 T/1460 C	2700 T
7	2000 T	7200 T	3600 T	1310 C	10800 T
8	970 C	3492 C	1746 C	1590 C/2830 T	5238 C
9	1302 C	4688 C	2344 C	2130 C/3800 T	7031 C

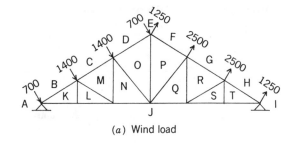

(a) Wind load

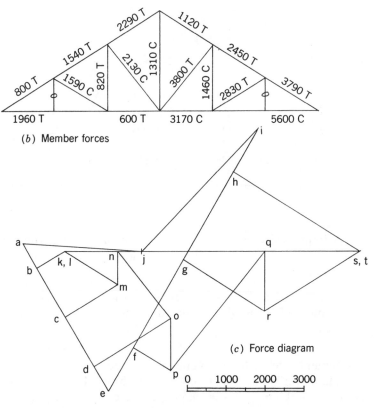

(b) Member forces

(c) Force diagram

FIGURE 12.38 Investigation of the roof truss for wind load.

TABLE 12.7 Design of the Truss Members

No.	Truss Member Force (kips)	Length (ft)	Member Choice (All Double-Angles)
1	26.3 C	11.7	Combined bending and compression member $6 \times 4 \times \frac{1}{2}$
2	21 C	11.7	Max. $L/r = 200$, min. $r = 0.7$ $6 \times 4 \times \frac{1}{2}$
3	22.5 T	10	Max. $L/r = 240$, min. $r = 0.5$ $3 \times 2\frac{1}{2} \times \frac{3}{8}$
4	18 T	10	$3 \times 2\frac{1}{2} \times \frac{3}{8}$
5	0	6	$2\frac{1}{2} \times 2\frac{1}{2} \times \frac{3}{8}$
6	2.7 T	12	Max. $L/r = 240$, min. $r = 0.6$ $2\frac{1}{2} \times 2\frac{1}{2} \times \frac{3}{8}$
7	10.8 T	18	Max. $L/r = 240$, min. $r = 0.9$ $2\frac{1}{2} \times 2\frac{1}{2} \times \frac{3}{8}$
8	5.24 C	11.7	Max. $L/r = 200$, min. $r = 0.7$ $2\frac{1}{2} \times 2\frac{1}{2} \times \frac{3}{8}$
9	7.03 C	15.6	Max. $L/r = 200$, min. $r = 0.78$ $3 \times 2\frac{1}{2} \times \frac{3}{8}$

- Minimum sizes of angle pairs, based on maximum permitted L/r ratios (120 for compression members and 200 for tension members).

- Layout of truss joints to avoid eccentricity of forces at the joints and to provide sufficient dimension for necessary welds.

After tentatively selecting members, designers must draw joints to scale to verify the construction's feasibility. If this practice exposes a problem, you well may need to reconsider some of the basic system. Possibly you need a closer spacing of trusses to reduce the truss loading, or maybe you must reconsider the truss member shapes or joint forms, or perhaps you ought to use an adapted manufactured truss.

Design for Lateral Forces

Given the construction shown in Figure 12.34, the most likely vertical bracing system is the exterior structural masonry walls. The
steel floor and roof construction, meanwhile, can be used for
horizontal diaphragm functions. While the sloping roof deck
might be used for diaphragm purpose, the most effective such bracing will use horizontal-plane trussing at the level of the trusses' bottom chords.

The second- and third-story floor structures consist of donut-
shaped surfaces (see Figure 12.39). These structures' behavior as
horizontal diaphragms, depends on the size of their holes (Figure
12.40 shows the extreme cases). A possible design approach for this
horizontal structure is as follows:

1. Design for the diaphragm shear in the deck, using the net
   width at each location for the shear stress investigation.
2. Design the individual subdiaphragms for their independent
   actions (see Figure 12.39).
3. Investigate for the chord forces in the whole diaphragm and
   the subdiaphragms to ensure that the steel framing can develop the forces.

Similar to roof trusses, the horizontal structure must work first
for the gravity loads. If lateral forces are minor (as in this example),
they are not critical for design, except to ensure that the construction includes the necessary anchors and maintains a framing continuity to perform chord, collector, tie, and drag strut functions.

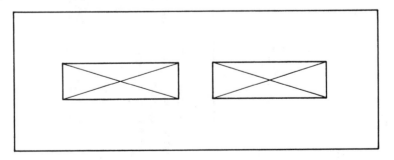

FIGURE 12.39 Schematic plan of the upper-level horizontal floor diaphragm.

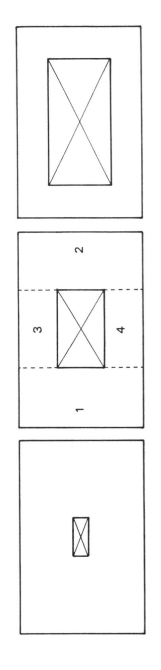

Small Opening:
functioning of whole diaphragm mainly unaffected; reinforce edges and corners of hole

Large Hole:
diaphragm reduced to parts (subdiaphragms) that may work as a connected set

Very Large Hole:
not a diaphragm, can function only as a very stiff rigid frame

FIGURE 12.40 Range of effects of a hole in a horizontal diaphragm. (a) Small hole with negligible effect on the diaphragm as a whole; requires only some reinforcement for the hole edges and corners. (b) Large hole, possibly indicating need for design of subdiaphragms (1, 2, 3, 4, as noted) and major reinforcement of the hole edges. (c) Hole so large that the diaphragm is incapable of effective action; may function as a rigid frame with additional vertical bracing.

12.7 BUILDING FIVE

Building Five is an open-sided, canopy roof structure (see Figure 12.41). Vertical support for the roof is provided by eight columns, each of which is developed as a built-up steel member formed of three steel plates welded together with fillet welds. The columns are joined to the gabled roof edge members and the corner canti-lever.

Eight T shaped edge frame units (see Figure 12.42) form the four framed bents that provide both vertical support for the rest of the roof construction and lateral bracing for the building. Essential to the latter function is the moment-resisting joint between the bent

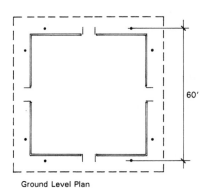

Ground Level Plan

60'

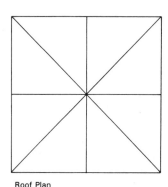

Roof Plan

Elevation

Framing Elements

1 T-bent with sloping top.
2 Gabled diagonal, sloped, main spanning member, forms roof valleys.
3 Horizontal ridge member.
4, 5 Horizontal beams, support deck.
6 Formed sheet steel deck.

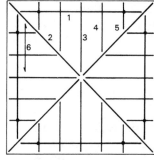

Roof Framing Plan

FIGURE 12.41 Building Five: General form.

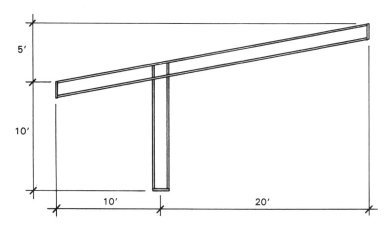

FIGURE 12.42 Building Five: Individual rigid frame unit.

column and the gabled member. These bents are very similar to those shown for the gabled roof example for Building One (see Figure 12.9).

The roof's interior portion is developed with the framing shown in Figure 12.41*d*. The gabled diagonals (designated No. 2 in the plan) span approximately 42 ft from corner to center and rise 10 ft. These simple-span elements carry concentrated loads consisting of the end reactions from members 4 and 5. The horizontal thrust at the corner is resisted by tension in the edge framing members.

Members 3, 4, and 5 are simple beams with a generally uniform load, modified to some degree by the triangular load areas. Although you must design some special connections for this structure, they are not uncommon; in fact, you can design them easily with special elements made from bent and welded steel plates.

12.8 BUILDING SIX

In this section I present some possibilities for developing the roof structure and exterior walls for a medium-size sports arena, one big enough for a natatorium or a basketball court (see Figure 12.43). Structural options are linked to the desired building form.

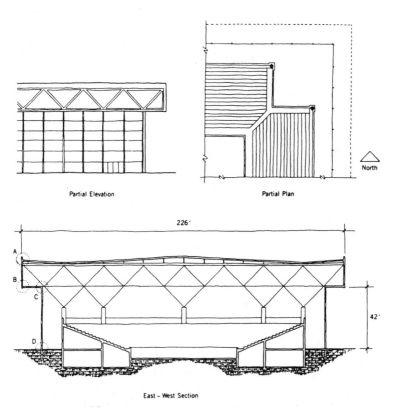

Partial Elevation   Partial Plan

△
North

226'

A

B

C

42'

D

East – West Section

FIGURE 12.43   Building Six: General form.

General Design Requirements

Functional planning *requirements* derive both from the specific activities to be housed and from the seating, internal traffic, overhead clearance, exits, and entrances. Despite all the requirements, designers usually can choose from a range of alternatives for the general building plan and overall form. For example, the truss system I discuss in this section is based on a commitment to a square plan and a flat roof profile. But a domed roof, on the other hand, would require a round plan.

Developing a 42-ft-high curtain wall is a major problem. Braced laterally only at the top and bottom, such a wall is a spanning structure that must sustain wind pressure as a loading. Even modest wind conditions—that is, producing pressures in the 20 psf range—require some major vertical mullion structure to span the 42 ft.

In this example the fascia of the roof trusses, the soffit of the overhang, and the curtain wall are all developed with products ordinarily available for curtain wall construction. The tall wall's vertical span is developed by custom-designed trusses that brace custom-developed major vertical mullions.

The truss system is an exposed structure, a major visual element of the building interior. As such, truss members define an ordered pattern. Meanwhile, within or beneath the trusses are many other items. For example,

- Elements of the roof drainage system, including leaders
- Ducts and registers for the HVAC system
- A general lighting system
- Signs, scoreboards, etc.
- Elements of an audio system
- Catwalks

To preserve some design order, designers try to relate such items to the truss geometry and detailing. However, you can't always match items installed or modified after the building is completed.

Structural Alternatives

The general construction form for Building Six is shown in Figure 12.44. I limit my discussion to the tall exterior glazed wall and the roof structure.

The exterior wall details in Figure 12.45 indicate the use of a standard, proprietary window wall system for the wall surface development. They also indicate the use of a custom-designed steel support structure; occurring behind the glazed wall and inside the building, the steel support structure is thus weather-protected. *Note:* In fact, the glazed wall ordinarily is developed as a weather-resistive construction system.

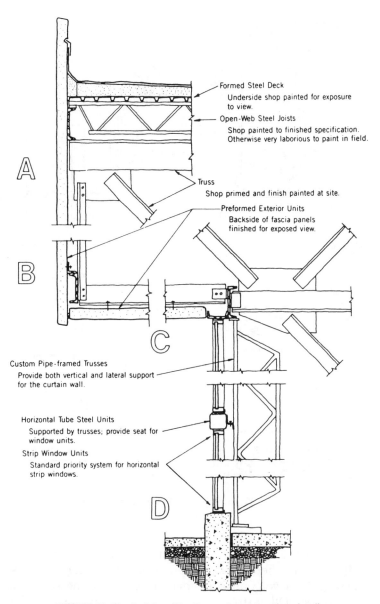

A

Formed Steel Deck
  Underside shop painted for exposure
  to view.
Open-Web Steel Joists
  Shop painted to finished specification.
  Otherwise very laborious to paint in field.

Truss
  Shop primed and finish painted at site.

Preformed Exterior Units
  Backside of fascia panels
  finished for exposed view.

B

C

Custom Pipe-framed Trusses
Provide both vertical and lateral support
for the curtain wall.

Horizontal Tube Steel Units
  Supported by trusses; provide seat for
  window units.
Strip Window Units
  Standard priority system for horizontal
  strip windows.

D

FIGURE 12.44 Building Six: General construction details.

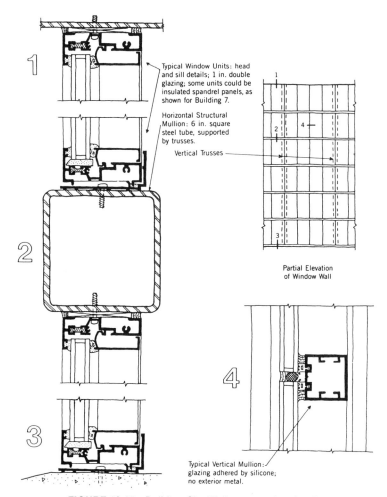

Typical Window Units: head and sill details; 1 in. double glazing; some units could be insulated spandrel panels, as shown for Building 7.

Horizontal Structural Mullion: 6 in. square steel tube, supported by trusses.

Vertical Trusses

Partial Elevation of Window Wall

Typical Vertical Mullion: glazing adhered by silicone; no exterior metal.

FIGURE 12.45 Building Six: Wall construction details.

The 168 ft clear span is definitely a long-span structure, but you still have a number of options. A flat-spanning beam system cannot work, but if you want a flat span, you can use a one- or two-way truss system. Other structural options, which involve some form other than a flat profile, include domes, arches, shells, folded plates, suspended cables, cable-stayed systems, and pneumatic systems.

The structure in Figure 12.43 uses a two-way spanning steel truss system; the form is called an *offset grid* (see Section 10.8). When planning this type structure, you must develop a module relating to the frequency of nodal points (i.e., joints) in the truss system. Truss supports go at nodal points, where concentrated loads should be applied. The nodal point module most affects the truss system, but designers can extend it to other aspects of the building planning. At the extreme, you may use the module throughout the building—as in this example.

The basic Building Six module is 3.5 ft (i.e., 42 in.). Multiples and fractions of this basic dimension appear throughout the building, in two and three dimensions. For example, the truss nodal module is 8X or 28 ft, the exterior wall height (ground to underside of truss) is 12X or 42 ft, and so on.

Building Six is not an ordinary building, although many examples exist. Almost always, designing such a building requires some innovation, unless you choose to duplicate some existing structure. Although Building Six is a straightforward example, not even the construction illustrated in this section can be called common or standard.

Even in unique buildings, however, designers try to use as many standard products as possible. In this case the roof structure on top of the truss and the curtain wall system for the exterior walls use off-the-shelf products (see Figures 12.44 and 12.45). The supporting columns and general seating structure also may consist of conventional construction.

The Window Wall Support Structure

The window wall system consists of a metal frame assembled from modular units with glazing inserted in the framed openings. Designed to support the glazing, the frame is self-supporting to a degree. When used for ordinary multistory buildings, this type frame can span the clear height of a single story (12 to 15 ft), although you may want to add some slightly stronger vertical mullions.

Because Building Six is about three times as tall as a normal single story, the wall requires some additional support to resist both the gravity load and lateral forces (especially wind, because the wall is quite light). Figures 12.44 and 12.45 illustrate the use of such a support system, consisting of trusses that span the 42 ft height and are spaced 14 ft on center and horizontal steel tubes that span the 14 ft between trusses and support 7 ft of the wall's vertical height.

Because the wall is relatively light, the vertical loads are carried by the closely spaced vertical mullions. In this case, the primary function of the horizontal steel tubes is to span between trusses and resist the wind forces on 14 ft of the wall. The tube shown is certainly adequate, although stiffness is probably more critical than bending stress in this situation: the wall construction must not flip-flop during earthquakes or when wind direction changes quickly.

The truss form is probably a *delta truss*, with two chords opposing a single one (creating a triangular cross section, like the Greek capital letter delta). This truss form boasts high lateral stability, so you can use it without cross-bridging.

The Truss System

When designing a structure such as Building Six, designers try first to use a pre-existing truss system. Custom designing such a structure requires a great deal of time—you must develop and plan the structural form, develop nodal joint construction, and investigate the structural behavior of the highly indeterminate structure. Moreover, assembling and erecting the truss pose other major design problems.

If the project deserves such effort, and if the time and money are available, the result may justify the expenditures. Otherwise, save some time and money and use available products and systems.

The square plan and general biaxial plan symmetry of Building Six indicate the need for a two-way spanning system. The truss system form illustrated in Figure 12.43 is called an *offset grid*, a name that refers to how the squares of the top chords are offset from those of the bottom chords so that the top chord nodules (joints) lie over the center of the bottom squares. As a result, there are no vertical web members or vertical planar sets in the system.

Note: I discussed two-way trusses in Section 10.8.

To achieve the general building form as shown in Figure 12.30, you can choose among a variety of roof truss systems. In this section I discuss several such options; often, cost is the most important factor governing a designer's decision.

Development of the Roof Infill System. The roof truss schemes I discuss involve providing a grid of truss chords at 28 ft centers in both directions at the roof surface (atop the trusses). This grid provides a roof support system, but not a roof, which requires an infill surface-developing construction whose top surface is water-resistant.

Building Six does not require an imaginative two-way spanning system; you can use simpler systems that fill in the 28 ft span voids. Figure 12.46, for example, shows a very simple system, consisting of simple-span steel open-web joists and formed sheet steel decking. In the simplest construction, the joists use only the truss chords in one direction for support; except for the related connection details, such an unsymmetrical loading for the two-way system does not affect how you develop the larger truss system.

Although the steel deck can be exposed on its underside, along with the open-web joists, you may want to provide a ceiling surface at the bottom of the open-web joists for extra thermal insulation and sound control.

If the open-web joists are placed as shown in Figure 12.46 (the usual installation), you may need to slope the two-way truss system top chords to achieve roof drainage. In fact, you may want to add some elements between the two-way trusses and the roof infill structure, permitting the two-way system to be flat on top, simplifying the truss system.

Note: This basic infill system is reasonable for the following alternatives.

Alternative One: The Offset Grid System. Figures 12.47 and 12.48 illustrate an offset grid system; note how the grid system is placed to provide supports beneath top chord joints.

The drawings in Figure 12.43 indicate the use of three columns on each side of the structure—a total of 12 supports for the truss system. The tops of the columns are dropped below the spanning truss to permit the use of a pyramidal module of four struts between the top of the column and the bottom chords of the truss system, reducing the maximum shear required by the truss interior members; in fact, the entire gravity load is shared by 48 truss members. If the total design gravity load is approximately 100 psf (0.1 kips/ft^2), the load in a single diagonal column strut is

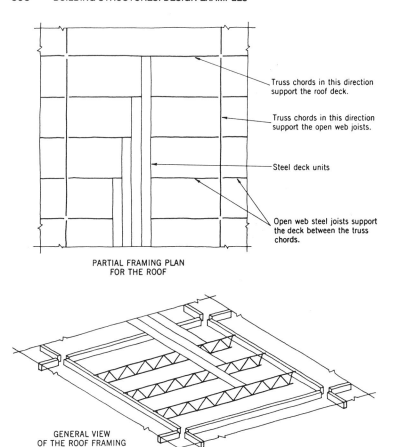

PARTIAL FRAMING PLAN
FOR THE ROOF

Truss chords in this direction
support the roof deck.

Truss chords in this direction
support the open web joists.

Steel deck units

Open web steel joists support
the deck between the truss
chords.

GENERAL VIEW
OF THE ROOF FRAMING

FIGURE 12.46 Building Six: Roof infill structure.

$$C = \frac{(226)^2(0.1)}{48} = 106 \text{ kips}$$

Since each strut picks up a truss node with four interior diagonals, the maximum internal force in the interior diagonals is

$$C = \frac{106}{4} = 26 \text{ kips}$$

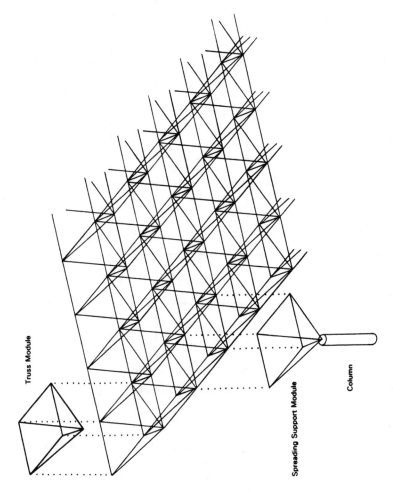

Truss Module

Spreading Support Module

Column

FIGURE 12.47   Building Six: General form of the offset grid system.

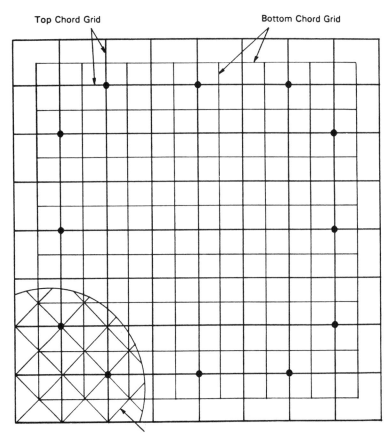

Top Chord Grid

Bottom Chord Grid

Plan of Truss Web Members

FIGURE 12.48 Plan layout of the offset grid system.

If you use steel pipe for the 28-ft-long members, you can choose 6 in. standard for the truss member (37 kip capacity) and 8 in. extra strong for the strut (137 kip capacity). For other choices, refer to tables in the AISC Manual (Ref. 3).

The closely spaced edge columns and the struts constitute almost a continuous edge support for the truss system, with only a minor edge cantilever. Thus the spanning task is essentially that of a simple beam span in two directions. For design purposes, consider the span in each direction to carry half the load. As a result, taking half the clear span width (168 ft) as a middle strip, the total load for design of the "simple beam" is

$$W = (span\ width)(span\ length)(0.1\ kips/ft^2)$$

$$= (84)(168)(0.1) = 1411\ kips$$

The simple beam moment at midspan is

$$M = \frac{WL}{8} = \frac{(1411)(168)}{8} = 29,631\ kip\text{-}ft$$

If there are three top chord members in the middle strip, and given a center-to-center chord depth of approximately 19 ft, the force in a single chord is

$$C = \frac{29631}{(3)(19)} = 520\ kips$$

If the compression chord is fully unbraced, such force exceeds the capacity of a pipe (at least from the AISC Tables) but not a W shape (W 12 × 136 with F_y of 50 ksi) or a pair of thick angles (8 × 8 × 1 in. with F_y = 50 ksi).

Developing joints at nodes or splices is critical for the bottom chords. Such chord members are rarely longer than a single module (28 ft), so each joint must be fully developed with welds or bolts.

Such sizes are not rare for large steel structures, but you should try to reduce the design loads (in other words, use the lightest possible general roof construction). You also may want to vary truss member sizes, using smaller members for low-stress situations.

As I discussed elsewhere in this book, this is a highly indeterminate but symmetrical structure. However, most professional structural designers can handle such a design, thanks to the availability of computer-aided procedures for investigation.

Alternative Two: The Two-Way Vertical Planar Truss System.
A second possible truss form consists of perpendicular, intersecting sets of vertical planar trusses (see Figure 12.49). In this system

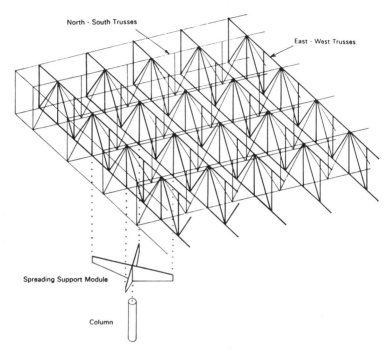

North - South Trusses

East - West Trusses

Spreading Support Module

Column

FIGURE 12.49 General form of the two-way truss system with vertical planar trusses.

the top chord grid squares sit directly above the bottom chord grid squares.

In Figure 12.50 the vertical truss planes are offset from the columns to permit the use of the spread unit at the column. This unit does not relate to the basic truss system form as it did for Alternative One, so it can take various shapes.

When calculating chord forces, you can use interior vertical members to reduce the lateral support problem. Using the same force approximations as determined for the chords in the preceding example, but assuming unsupported lengths of only 14 ft, you can design for considerably smaller members.

For both two-way systems, you must plan for the erection by identifying what size and shape unit can be assembled in the shop and transported to the site. You also must determine what temporary support is needed. Such details may affect how you design the truss jointing details.

Edge Carrying Trusses   Two-Way Center Truss Grid System

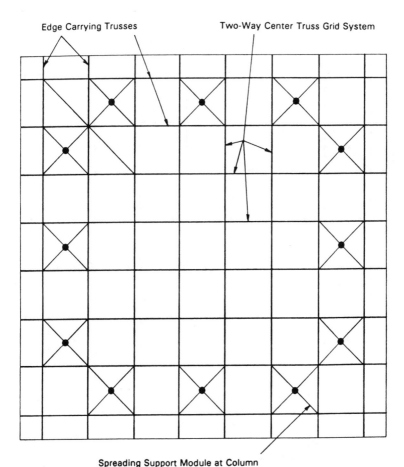

Spreading Support Module at Column

FIGURE 12.50   Plan layout of the system with vertical planar trusses.

Alternative Three: The One-Way System.   Figure 12.51 shows
a system that uses a set of one-way spanning, planar trusses. The
general truss form is identical to the system for Alternative Two but
the manner in which trusses are formed and joints are achieved
is different.

In this example, the span is achieved by the set of trusses spanning in one direction; the cross-trussing is used only to span between the carrying trusses and to provide lateral bracing for the
system. The cross-trussing also cantilevers to develop the facia and
soffit on two sides of the building.

Primary Carrying Trusses   Secondary Carrying Trusses

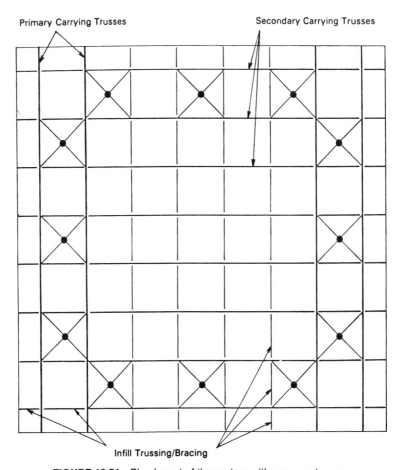

Infill Trussing/Bracing

FIGURE 12.51 Plan layout of the system with one-way trusses.

This system is simple to design because the main trusses are simple, planar, determinate trusses. This simplicity should not compel you to choose this scheme, but it is food for thought, given the complexity of investigations of highly indeterminate systems.

The carrying trusses are slightly heavier than the trusses in Alternative Two because their share of the load is slightly more, but the cross-trussing system's structural demands are minor.

The designer must decide how far to go to make the structure appear symmetrical in an otherwise biaxially symmetrical building. In reality, however, most nonprofessionals never notice a lack of symmetry. In fact, most of our design subtlety is typicallly lost on all but our fellow professionals.

The best reason to favor this scheme: It is simple to assemble and erect. You may erect carrying trusses in one piece, with very little temporary support. Once two that straddle a column are in place, you can develop the cross-trussing, which serves both temporary and permanent bracing functions.

_____

PROPERTIES OF SHAPES

This appendix contains data for the properties of cross sections of structural members.

A.1 PROPERTIES OF COMMON GEOMETRIC SHAPES

Figure A.1 gives formulas for various geometric forms. Although more common in wood or concrete, solid cross sections are sometimes used in steel. Many designers form fabricated members with open or hollow cross sections by bending and welding steel elements, such as plates and angles.

A.2 PROPERTIES OF ROLLED STEEL SHAPES

The most common structural elements are formed by hot-rolling, which I described in Section 2.3. Tables A.1 through A.3 contain data for selected shapes. Table A.4 contains properties of steel pipe.

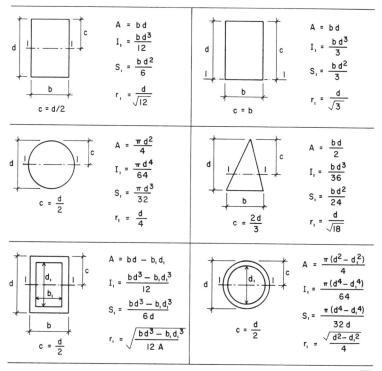

A = Area I = Moment of inertia S = Section modulus = $\frac{I}{c}$ r = Radius of gyration = $\sqrt{\frac{I}{A}}$

FIGURE A.1 Properties of common geometric shapes.

This data is adapted from more extensive tables in the AISC Manual with permission of the publishers, American Institute of Steel Construction. I provide this data for use in example computations. In addition, I offer these tables so that readers may become familiar with the kinds of tables found in the AISC Manual.

TABLE A.1 Properties of W Shapes

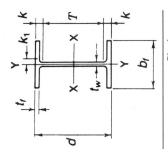

| | Identification | | | Dimensions | | | | | Elastic Properties | | | | | | Plastic Modulus | |
| | | | | | | | | | Axis X-X | | | Axis Y-Y | | | | |
Designation	Area A (in.²)	Depth d (in.)	t_w (in.)	b_f (in.)	t_f (in.)	k (in.)	k_1 (in.)	I (in.⁴)	S (in.³)	r (in.)	I (in.⁴)	S (in.³)	r (in.)	Z_x (in.³)	Z_y (in.³)
W 30×116	34.2	30.01	0.565	10.495	0.850	1⁵/₈	1	4930	329	12.0	164	31.3	2.19	378	49.2
×108	31.7	29.83	0.545	10.475	0.760	1⁹/₁₆	1	4470	299	11.9	146	27.9	2.15	346	43.9
× 99	29.1	29.65	0.520	10.450	0.670	1⁷/₁₆	1	3990	269	11.7	128	24.5	2.10	312	38.6
W 27× 94	27.7	26.92	0.490	9.990	0.745	1⁷/₁₆	¹⁵/₁₆	3270	243	10.9	124	24.8	2.12	278	38.8
× 84	24.8	26.71	0.460	9.960	0.640	1³/₈	¹⁵/₁₆	2850	213	10.7	106	21.2	2.07	244	33.2
W 24× 84	24.7	24.10	0.470	9.020	0.770	1⁹/₁₆	¹⁵/₁₆	2370	196	9.79	94.4	20.9	1.95	224	32.6
× 76	22.4	23.92	0.440	8.990	0.680	1⁷/₁₆	¹⁵/₁₆	2100	176	9.69	82.5	18.4	1.92	200	28.6
× 68	20.1	23.73	0.415	8.965	0.585	1³/₈	¹⁵/₁₆	1830	154	9.55	70.4	15.7	1.87	177	24.5
W 21× 83	24.3	21.43	0.515	8.355	0.835	1⁹/₁₆	¹⁵/₁₆	1830	171	8.67	81.4	19.5	1.83	196	30.5
× 73	21.5	21.24	0.455	8.295	0.740	1¹/₂	¹⁵/₁₆	1600	151	8.64	70.6	17.0	1.81	172	26.6

(Continued)

TABLE A.1 Continued

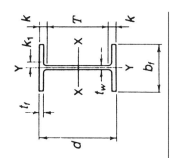

Identification			Dimensions					Elastic Properties						Plastic Modulus	
								Axis X-X			Axis Y-Y				
Designation	Area A (in.²)	Depth d (in.)	t_w (in.)	b_f (in.)	t_f (in.)	k (in.)	k_1 (in.)	I (in.⁴)	S (in.³)	r (in.)	I (in.⁴)	S (in.³)	r (in.)	Z_x (in.³)	Z_y (in.³)
W 21× 57	16.7	21.06	0.405	6.555	0.650	1 3/8	7/8	1170	111	8.36	30.6	9.35	1.35	129	14.8
× 50	14.7	20.83	0.380	6.530	0.535	1 5/16	7/8	984	94.5	8.18	24.9	7.64	1.30	110	12.2
W 18× 86	25.3	18.39	0.480	11.090	0.770	1 7/16	7/8	1530	166	7.77	175	31.6	2.63	186	48.4
× 76	22.3	18.21	0.425	11.035	0.680	1 3/8	13/16	1330	146	7.73	152	27.6	2.61	163	42.2
W 18× 60	17.6	18.24	0.415	7.555	0.695	1 3/8	13/16	984	108	7.47	50.1	13.3	1.69	123	20.6
× 55	16.2	18.11	0.390	7.530	0.630	1 5/16	13/16	890	98.3	7.41	44.9	11.9	1.67	112	18.5
× 50	14.7	17.99	0.355	7.495	0.570	1 1/4	13/16	800	88.9	7.38	40.1	10.7	1.65	101	16.6
W 18× 46	13.5	18.06	0.360	6.060	0.605	1 1/4	13/16	712	78.8	7.25	22.5	7.43	1.29	90.7	11.7
× 40	11.8	17.90	0.315	6.015	0.525	1 3/16	13/16	612	68.4	7.21	19.1	6.35	1.27	78.4	9.95
W 16× 50	14.7	16.26	0.380	7.070	0.630	1 5/16	13/16	659	81.0	6.68	37.2	10.5	1.59	92.0	16.3
× 45	13.3	16.13	0.345	7.035	0.565	1 1/4	13/16	586	72.7	6.65	32.8	9.34	1.57	82.3	14.5
× 40	11.8	16.01	0.305	6.995	0.505	1 3/16	13/16	518	64.7	6.63	28.9	8.25	1.57	72.9	12.7
× 36	10.6	15.86	0.295	6.985	0.430	1 1/8	3/4	448	56.5	6.51	24.5	7.00	1.52	64.0	10.8

W 14×216	62.0	15.72	0.980	15.800	1.560	2 1/4	1 1/8	2660	338	6.55	1030	130	4.07	390	198
×176	51.8	15.22	0.830	15.650	1.310	2	1 1/16	2140	281	6.43	838	107	4.02	320	163
W 14×132	38.8	14.66	0.645	14.725	1.030	1 11/16	15/16	1530	209	6.28	548	74.5	3.76	234	113
×120	35.3	14.48	0.590	14.670	0.940	1 5/8	15/16	1380	190	6.24	495	67.5	3.74	212	102
W 14× 74	21.8	14.17	0.450	10.070	0.785	1 9/16	15/16	796	112	6.04	134	26.6	2.48	126	40.6
× 68	20.0	14.04	0.415	10.035	0.720	1 1/2	15/16	723	103	6.01	121	24.2	2.46	115	36.9
W 14× 48	14.1	13.79	0.340	8.030	0.595	1 1/8	7/8	485	70.3	5.85	51.4	12.8	1.91	78.4	19.6
× 43	12.6	13.66	0.305	7.995	0.530	1 5/16	7/8	428	62.7	5.82	45.2	11.3	1.89	69.6	17.3
W 14× 34	10.0	13.98	0.285	6.745	0.455	1	5/8	340	48.6	5.83	23.3	6.91	1.53	54.6	10.6
× 30	8.85	13.84	0.270	6.730	0.385	15/16	5/8	291	42.0	5.73	19.6	5.82	1.49	47.3	8.99
W 12×136	39.9	13.41	0.790	12.400	1.250	1 15/16	1	1240	186	5.58	398	64.2	3.16	214	98.0
×120	35.3	13.12	0.710	12.320	1.105	1 13/16	1	1070	163	5.51	345	56.0	3.13	186	85.4
W 12× 72	21.1	12.25	0.430	12.040	0.670	1 3/8	7/8	597	97.4	5.31	195	32.4	3.04	108	49.2
× 65	19.1	12.12	0.390	12.000	0.605	1 5/16	13/16	533	87.9	5.28	174	29.1	3.02	96.8	44.1
W 12× 53	15.6	12.06	0.345	9.995	0.575	1 1/4	13/16	425	70.6	5.23	95.8	19.2	2.48	77.9	29.1
W 12× 45	13.2	12.06	0.335	8.045	0.575	1 1/4	13/16	350	58.1	5.15	50.0	12.4	1.94	64.7	19.0
× 40	11.8	11.94	0.295	8.005	0.515	1 1/4	3/4	310	51.9	5.13	44.1	11.0	1.93	57.5	16.8
W 12× 30	8.79	12.34	0.260	6.520	0.440	15/16	1/2	238	38.6	5.21	20.3	6.24	1.52	43.1	9.56
× 26	7.65	12.22	0.230	6.490	0.380	7/8	1/2	204	33.4	5.17	17.3	5.34	1.51	37.2	8.17
W 10× 88	25.9	10.84	0.605	10.265	0.990	1 1/8	13/16	534	98.5	4.54	179	34.8	2.63	113	53.1
× 77	22.6	10.60	0.530	10.190	0.870	1 1/2	13/16	455	85.9	4.49	154	30.1	2.60	97.6	45.9
× 49	14.4	9.98	0.340	10.000	0.560	1 9/16	11/16	272	54.6	4.35	93.4	18.7	2.54	60.4	28.3
W 10× 39	11.5	9.92	0.315	7.985	0.530	1 1/8	11/16	209	42.1	4.27	45.0	11.3	1.98	46.8	17.2
× 33	9.71	9.73	0.290	7.960	0.435	1 11/16	11/16	170	35.0	4.19	36.6	9.20	1.94	38.8	14.0
W 10× 19	5.62	10.24	0.250	4.020	0.395	13/16	1/2	96.3	18.8	4.14	4.29	2.14	0.874	21.6	3.35
× 17	4.99	10.11	0.240	4.010	0.330	3/4	1/2	81.9	16.2	4.05	3.56	1.78	0.844	18.7	2.80

Source: Adapted from data in the *Manual of Steel Construction*, 8th edition, with permission of the publishers, American Institute of Steel Construction. This table is a sample from an extensive set of tables.

TABLE A.2   Properties of Selected Single-Angle Shapes

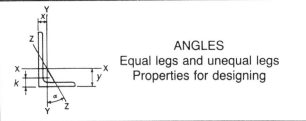

ANGLES
Equal legs and unequal legs
Properties for designing

Size and Thickness	k	Weight per Ft	Area	AXIS X-X				AXIS Y-Y				AXIS Z-Z	
				I	S	r	y	I	S	r	x	r	Tan
In.	In.	Lb.	In.^{2}	In.^{4}	In.^{3}	In.	In.	In.^{4}	In.^{3}	In.	In.	In.	α
L 8×8×1⅛	1¾	56.9	16.7	98.0	17.5	2.42	2.41	98.0	17.5	2.42	2.41	1.56	1.000
1	1⅝	51.0	15.0	89.0	15.8	2.44	2.37	89.0	15.8	2.44	2.37	1.56	1.000
L 8×6× ¾	1¼	33.8	9.94	63.4	11.7	2.53	2.56	30.7	6.92	1.76	1.56	1.29	0.551
½	1	23.0	6.75	44.3	8.02	2.56	2.47	21.7	4.79	1.79	1.47	1.30	0.558
L 6×6 × ⅝	1⅛	24.2	7.11	24.2	5.66	1.84	1.73	24.2	5.66	1.84	1.73	1.18	1.000
½	1	19.6	5.75	19.9	4.61	1.86	1.68	19.9	4.61	1.86	1.68	1.18	1.000
L 6×4 × ⅝	1⅛	20.0	5.86	21.1	5.31	1.90	2.03	7.52	2.54	1.13	1.03	0.864	0.435
½	1	16.2	4.75	17.4	4.33	1.91	1.99	6.27	2.08	1.15	0.987	0.870	0.440
⅜	⅞	12.3	3.61	13.5	3.32	1.93	1.94	4.90	1.60	1.17	0.941	0.877	0.446
L 5×3½× ½	1	13.6	4.00	9.99	2.99	1.58	1.66	4.05	1.56	1.01	0.906	0.755	0.479
⅜	⅞	10.4	3.05	7.78	2.29	1.60	1.61	3.18	1.21	1.02	0.861	0.762	0.486
L 5×3 × ½	1	12.8	3.75	9.45	2.91	1.59	1.75	2.58	1.15	0.829	0.750	0.648	0.357
⅜	⅞	9.8	2.86	7.37	2.24	1.61	1.70	2.04	0.888	0.845	0.704	0.654	0.364
L 4×4 × ½	⅞	12.8	3.75	5.56	1.97	1.22	1.18	5.56	1.97	1.22	1.18	0.782	1.000
⅜	¾	9.8	2.86	4.36	1.52	1.23	1.14	4.36	1.52	1.23	1.14	0.788	1.000
L 4 ×3 × ½	15⁄16	11.1	3.25	5.05	1.89	1.25	1.33	2.42	1.12	0.864	0.827	0.639	0.543
⅜	13⁄16	8.5	2.48	3.96	1.46	1.26	1.28	1.92	0.866	0.879	0.782	0.644	0.551
5⁄16	¾	7.2	2.09	3.38	1.23	1.27	1.26	1.65	0.734	0.887	0.759	0.647	0.554
L 3½×3½× ⅜	¾	8.5	2.48	2.87	1.15	1.07	1.01	2.87	1.15	1.07	1.01	0.687	1.000
5⁄16	11⁄16	7.2	2.09	2.45	0.976	1.08	0.990	2.45	0.976	1.08	0.990	0.690	1.000
L 3½×2½× ⅜	13⁄16	7.2	2.11	2.56	1.09	1.10	1.16	1.09	0.592	0.719	0.660	0.537	0.496
5⁄16	¾	6.1	1.78	2.19	0.927	1.11	1.14	0.939	0.504	0.727	0.637	0.540	0.501
L 3 ×3 × ⅜	11⁄16	7.2	2.11	1.76	0.833	0.913	0.888	1.76	0.833	0.913	0.888	0.587	1.000
5⁄16	⅝	6.1	1.78	1.51	0.707	0.922	0.865	1.51	0.707	0.922	0.865	0.589	1.000
L 3 ×2½× ⅜	¾	6.6	1.92	1.66	0.810	0.928	0.956	1.04	0.581	0.736	0.706	0.522	0.676
5⁄16	11⁄16	5.6	1.62	1.42	0.688	0.937	0.933	0.898	0.494	0.744	0.683	0.525	0.680
L 3 ×2 × ⅜	11⁄16	5.9	1.73	1.53	0.781	0.940	1.04	0.543	0.371	0.559	0.539	0.430	0.428
5⁄16	⅝	5.0	1.46	1.32	0.664	0.948	1.02	0.470	0.317	0.567	0.516	0.432	0.435
L 2½×2½× ⅜	11⁄16	5.9	1.73	0.984	0.566	0.753	0.762	0.984	0.566	0.753	0.762	0.487	1.000
5⁄16	⅝	5.0	1.46	0.849	0.482	0.761	0.740	0.849	0.482	0.761	0.740	0.489	1.000
L 2½×2 × ⅜	11⁄16	5.3	1.55	0.912	0.547	0.768	0.831	0.514	0.363	0.577	0.581	0.420	0.614
5⁄16	⅝	4.5	1.31	0.788	0.466	0.776	0.809	0.446	0.310	0.584	0.559	0.422	0.620

Source: Adapted from the *Manual of Steel Construction*, 8th edition, with permission of the publisher, American Institute of Steel Construction. This table is a sample from an extensive set of tables.

TABLE A.3   Properties of Selected Double-Angle Shapes with Long Legs Back to Back

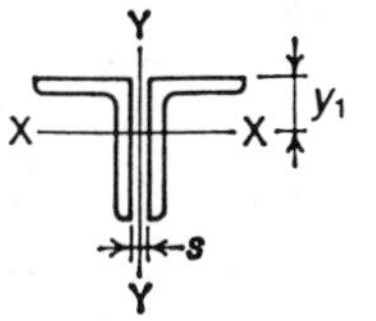

| | | | DOUBLE ANGLES
Two unequal leg angles
Properties of sections
Long legs back to back | | | | | | |

Designation	Wt. per Ft 2 Angles	Area of 2 Angles	AXIS X-X				AXIS Y-Y		
			I	S	r	y	Radii of Gyration Back to Back of Angles, In.		
	Lb.	In.^{2}	In.^{4}	In.^{3}	In.	In.	0	⅜	¾
L 8×6 ×1	88.4	26.0	161.0	30.2	2.49	2.65	2.39	2.52	2.66
¾	67.6	19.9	126.0	23.3	2.53	2.56	2.35	2.48	2.62
½	46.0	13.5	88.6	16.0	2.56	2.47	2.32	2.44	2.57
L 6×4 × ¾	47.2	13.9	49.0	12.5	1.88	2.08	1.55	1.69	1.83
⅝	40.0	11.7	42.1	10.6	1.90	2.03	1.53	1.67	1.81
½	32.4	9.50	34.8	8.67	1.91	1.99	1.51	1.64	1.78
⅜	24.6	7.22	26.9	6.64	1.93	1.94	1.50	1.62	1.76
L 5×3½× ¾	39.6	11.6	27.8	8.55	1.55	1.75	1.40	1.53	1.68
½	27.2	8.00	20.0	5.97	1.58	1.66	1.35	1.49	1.63
⅜	20.8	6.09	15.6	4.59	1.60	1.61	1.34	1.46	1.60
L 5×3 × ½	25.6	7.50	18.9	5.82	1.59	1.75	1.12	1.25	1.40
⅜	19.6	5.72	14.7	4.47	1.61	1.70	1.10	1.23	1.37
5⁄16	16.4	4.80	12.5	3.77	1.61	1.68	1.09	1.22	1.36
L 4 ×3 ×½	22.2	6.50	10.1	3.78	1.25	1.33	1.20	1.33	1.48
⅜	17.0	4.97	7.93	2.92	1.26	1.28	1.18	1.31	1.45
5⁄16	14.4	4.18	6.76	2.47	1.27	1.26	1.17	1.30	1.44
L 3½×2½×⅜	14.4	4.22	5.12	2.19	1.10	1.16	.976	1.11	1.26
5⁄16	12.2	3.55	4.38	1.85	1.11	1.14	.966	1.10	1.25
¼	9.8	2.88	3.60	1.51	1.12	1.11	.958	1.09	1.23
L 3 ×2 ×⅜	11.8	3.47	3.06	1.56	.940	1.04	.777	.917	1.07
5⁄16	10.0	2.93	2.63	1.33	.948	1.02	.767	.903	1.06
¼	8.2	2.38	2.17	1.08	.957	.993	.757	.891	1.04
L 2½×2 ×⅜	10.6	3.09	1.82	1.09	.768	.831	.819	.961	1.12
5⁄16	9.0	2.62	1.58	.932	.776	.809	.809	.948	1.10
¼	7.2	2.13	1.31	.763	.784	.787	.799	.935	1.09

Source: Adapted from the *Manual of Steel Construction*, 8th edition, with permission of the publisher, American Institute of Steel Construction. This table is a sample from an extensive set of tables.

TABLE A.4   Properties of Standard-Weight Steel Pipe

<table>
<tr><td colspan="10" align="center"># PIPE
Dimensions and properties</td></tr>
</table>

	Dimensions			Weight	Properties				
Nominal Diameter In.	Outside Diameter In.	Inside Diameter In.	Wall Thickness In.	per Ft Lbs. Plain Ends	A In.^{2}	I In.^{4}	S In.^{3}	r In.	Schedule No.
				Standard Weight					
½	.840	.622	.109	.85	.250	.017	.041	.261	40
¾	1.050	.824	.113	1.13	.333	.037	.071	.334	40
1	1.315	1.049	.133	1.68	.494	.087	.133	.421	40
1¼	1.660	1.380	.140	2.27	.669	.195	.235	.540	40
1½	1.900	1.610	.145	2.72	.799	.310	.326	.623	40
2	2.375	2.067	.154	3.65	1.07	.666	.561	.787	40
2½	2.875	2.469	.203	5.79	1.70	1.53	1.06	.947	40
3	3.500	3.068	.216	7.58	2.23	3.02	1.72	1.16	40
3½	4.000	3.548	.226	9.11	2.68	4.79	2.39	1.34	40
4	4.500	4.026	.237	10.79	3.17	7.23	3.21	1.51	40
5	5.563	5.047	.258	14.62	4.30	15.2	5.45	1.88	40
6	6.625	6.065	.280	18.97	5.58	28.1	8.50	2.25	40
8	8.625	7.981	.322	28.55	8.40	72.5	16.8	2.94	40
10	10.750	10.020	.365	40.48	11.9	161	29.9	3.67	40
12	12.750	12.000	.375	49.56	14.6	279	43.8	4.38	—

The listed sections are available in conformance with ASTM Specification A53 Grade B or A501. Other sections are made to these specifications. Consult with pipe manufacturers or distributors for availability.

Note: Pipe is also available in two heavier weights.

Source: Adapted from the *Manual of Steel Construction*, 8th edition, with permission of the publisher, American Institute of Steel Construction. This table is a sample from an extensive set of tables.

BEAM DESIGN AIDS

This appendix includes materials that help designers design steel beams.

Designers can use Table B.1 to rapidly select shapes for a maximum bending moment. Shapes are listed in the table in descending order of the magnitudes of their section modulus values for the primary bending axis: S_x for the x-x axis. The table also lists the safe service load bending moment corresponding to the S_x value and an allowable stress based on a yield stress of 36 ksi. (*Note:* The AISC Manual also has a table that gives bending moments based on a yield stress of 50 ksi.) Shapes are grouped in the table; each group is headed by the least weight member (listed in bold type)—that is, no other beam is as strong with less weight. The table also yields values for the lateral unsupported length limits of L_c and L_u.

From the two graphs shown in Figure B.1, you can determine safe beams for given combinations you can determine safe beams for given combinations of maximum bending moment and laterally unbraced length. I explained how to use these graphs in Section 6.5.

TABLE B.1   Allowable Stress Design Selection for Shapes Used as Beams

S_x	Shape	F_y = 36 ksi			S_x	Shape	F_y = 36 ksi		
		L_c	L_u	M_R			L_c	L_u	M_R
In.³		Ft.	Ft.	Kip-ft.	In.³		Ft.	Ft.	Kip-ft.
1110	W 36x300	17.6	35.3	2220	269	W 30x 99	10.9	11.4	538
1030	W 36x280	17.5	33.1	2060	267	W 27x102	10.6	14.2	534
953	W 36x260	17.5	30.5	1910	258	W 24x104	13.5	18.4	516
895	W 36x245	17.4	28.6	1790	249	W 21x111	13.0	23.3	498
837	W 36x230	17.4	26.8	1670	243	W 27x 94	10.5	12.8	486
829	W 33x241	16.7	30.1	1660	231	W 18x119	11.9	29.1	462
757	W 33x221	16.7	27.6	1510	227	W 21x101	13.0	21.3	454
719	W 36x210	12.9	20.9	1440	222	W 24x 94	9.6	15.1	444
684	W 33x201	16.6	24.9	1370	213	W 27x 84	10.5	11.0	426
664	W 36x194	12.8	19.4	1330	204	W 18x106	11.8	26.0	408
663	W 30x211	15.9	29.7	1330	196	W 24x 84	9.5	13.3	392
623	W 36x182	12.7	18.2	1250	192	W 21x 93	8.9	16.8	384
598	W 30x191	15.9	26.9	1200	190	W 14x120	15.5	44.1	380
580	W 36x170	12.7	17.0	1160	188	W 18x 97	11.8	24.1	376
542	W 36x160	12.7	15.7	1080	176	W 24x 76	9.5	11.8	352
539	W 30x173	15.8	24.2	1080	175	W 16x100	11.0	28.1	350
504	W 36x150	12.6	14.6	1010	173	W 14x109	15.4	40.6	346
502	W 27x178	14.9	27.9	1000	171	W 21x 83	8.8	15.1	342
487	W 33x152	12.2	16.9	974	166	W 18x 86	11.7	21.5	332
455	W 27x161	14.8	25.4	910	157	W14x 99	15.4	37.0	314
448	W 33x141	12.2	15.4	896	155	W 16x 89	10.9	25.0	310
439	W 36x135	12.3	13.0	878	154	W 24x 68	9.5	10.2	308
414	W 24x162	13.7	29.3	828	151	W 21x 73	8.8	13.4	302
411	W 27x146	14.7	23.0	822	146	W 18x 76	11.6	19.1	292
406	W 33x130	12.1	13.8	812	143	W 14x 90	15.3	34.0	286
380	W 30x132	11.1	16.1	760	140	W 21x 68	8.7	12.4	280
371	W 24x146	13.6	26.3	742	134	W 16x 77	10.9	21.9	268
359	W 33x118	12.0	12.6	718	131	W 24x 62	7.4	8.1	262
355	W 30x124	11.1	15.0	710	127	W 21x 62	8.7	11.2	254
329	W 30x116	11.1	13.8	658	127	W 18x 71	8.1	15.5	.254
329	W 24x131	13.6	23.4	658	123	W 14x 82	10.7	28.1	246
329	W 21x147	13.2	30.3	658	118	W 12x 87	12.8	36.2	236
299	W 30x108	11.1	12.3	598	117	W 18x 65	8.0	14.4	234
299	W 27x114	10.6	15.9	598	117	W 16x 67	10.8	19.3	234
295	W 21x132	13.1	27.2	590	114	W 24x 55	7.0	7.5	228
291	W 24x117	13.5	20.8	582	112	W 14x 74	10.6	25.9	224
273	W 21x122	13.1	25.4	546	111	W 21x 57	6.9	9.4	222
					108	W 18x 60	8.0	13.3	216
					107	W 12x 79	12.8	33.3	214
					103	W 14x 68	10.6	23.9	206
					98.3	W 18x 55	7.9	12.1	197
					97.4	W 12x 72	12.7	30.5	195

416

TABLE B.1 *Continued*

S_x	Shape	$F_y = 36$ ksi				S_x	Shape	$F_y = 36$ ksi		
		L_c	L_u	M_R				L_c	L_u	M_R
In.^{3}		Ft.	Ft.	Kip-ft.		In.^{3}		Ft.	Ft.	Kip-ft.
94.5	W 21x50	6.9	7.8	189		29.0	W 14x22	5.3	5.6	58
92.2	W 16x57	7.5	14.3	184		27.9	W 10x26	6.1	11.4	56
92.2	W 14x61	10.6	21.5	184		27.5	W 8x31	8.4	20.1	55
88.9	W 18x50	7.9	11.0	178		25.4	W 12x22	4.3	6.4	51
87.9	W 12x65	12.7	27.7	176		24.3	W 8x28	6.9	17.5	49
81.6	W 21x44	6.6	7.0	163		23.2	W 10x22	6.1	9.4	46
81.0·	W 16x50	7.5	12.7	162						
78.8	W 18x46	6.4	9.4	158		21.3	W 12x19	4.2	5.3	43
78.0	W 12x58	10.6	24.4	156						
77.8	W 14x53	8.5	17.7	156		21.1	M 14x18	3.6	4.0	42
72.7	W 16x45	7.4	11.4	145		20.9	W 8x24	6.9	15.2	42
70.6	W 12x53	10.6	22.0	141		18.8	W 10x19	4.2	7.2	38
70.3	W 14x48	8.5	16.0	141		18.2	W 8x21	5.6	11.8	36
68.4	W 18x40	6.3	8.2	137		17.1	W 12x16	4.1	4.3	34
66.7	W 10x60	10.6	31.1	133		16.7	W 6x25	6.4	20.0	33
						16.2	W 10x17	4.2	6.1	32
64.7	W 16x40	7.4	10.2	129		15.2	W 8x18	5.5	9.9	30
64.7	W 12x50	8.5	19.6	129						
62.7	W 14x43	8.4	14.4	125		14.9	W 12x14	3.5	4.2	30
60.0	W 10x54	10.6	28.2	120		13.8	W 10x15	4.2	5.0	28
58.1	W 12x45	8.5	17.7	116		13.4	W 6x20	6.4	16.4	27
						13.0	M 6x20	6.3	17.4	26
57.6	W 18x35	6.3	6.7	115						
56.5	W 16x36	7.4	8.8	113		12.0	M 12x11.8	2.7	3.0	24
54.6	W 14x38	7.1	11.5	109		11.8	W 8x15	4.2	7.2	24
54.6	W 10x49	10.6	26.0	109		10.9	W 10x12	3.9	4.3	22
51.9	W 12x40	8.4	16.0	104		10.2	W 6x16	4.3	12.0	20
49.1	W 10x45	8.5	22.8	98		10.2	W 5x19	5.3	19.5	20
						9.91	W 8x13	4.2	5.9	20
48.6	W 14x34	7.1	10.2	97		9.72	W 6x15	6.3	12.0	19
						9.63	M 5x18.9	5.3	19.3	19
47.2	W 16x31	5.8	7.1	94		8.51	W 5x16	5.3	16.7	17
45.6	W 12x35	6.9	12.6	91						
42.1	W 10x39	8.4	19.8	84		7.81	W 8x10	4.2	4.7	16
42.0	W 14x30	7.1	8.7	84		7.76	M 10x 9	2.6	2.7	16
						7.31	W 6x12	4.2	8.6	15
38.6	W 12x30	6.9	10.8	77						
						5.56	W 6x 9	4.2	6.7	11
38.4	W 16x26	5.6	6.0	77		5.46	W 4x13	4.3	15.6	11
						5.24	M 4x13	4.2	16.9	10
35.3	W 14x26	5.3	7.0	71						
35.0	W 10x33	8.4	16.5	70		4.62	M 8x 6.5	2.4	2.5	9
33.4	W 12x26	6.9	9.4	67		2.40	M 6x 4.4	1.9	2.4	5
32.4	W 10x30	6.1	13.1	65						
31.2	W 8x35	8.5	22.6	62						

Source: Adapted from the *Manual of Steel Construction*, 8th edition, with permission of the publisher, American Institute of Steel Construction.

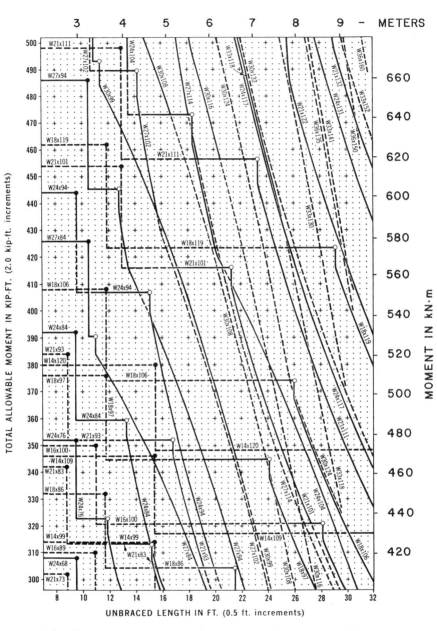

FIGURE B.1 Allowable moments in beams with various laterally unbraced lengths ($C_b = 1$, $F_y = 36$ ksi). Reproduced from *Manual of Steel Construction: Allowable Stress Design*, 8th edition, with permission of the publisher, American Institute of Steel Construction. This figure is a sample from a large set of tables.

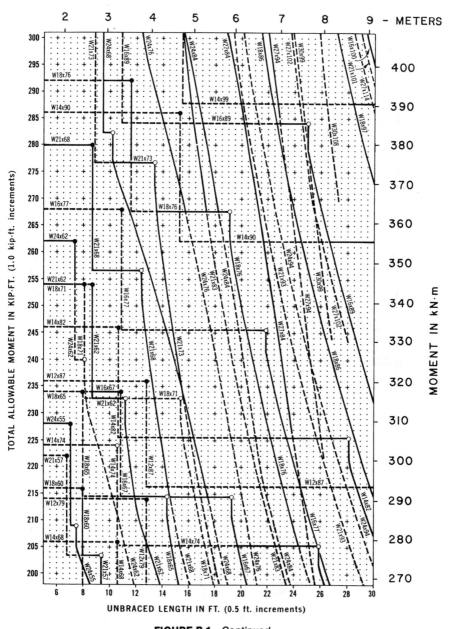

FIGURE B.1 *Continued*

419

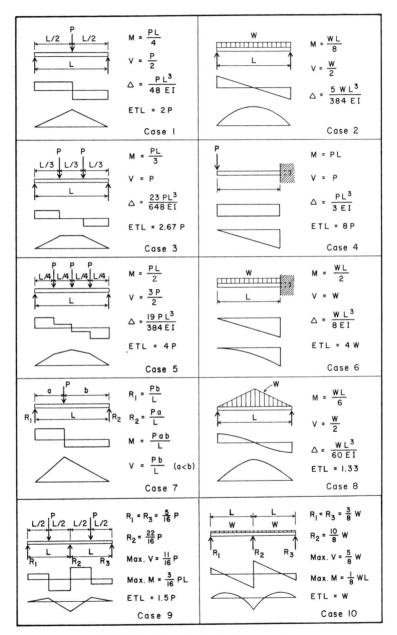

FIGURE B.2 Values for typical beam loadings.

Figure B.2 shows ten loading and support configurations for beams, with formulas for determination of values for reaction forces, maximum shear, maximum deflection, and, for some cases, maximum deflection. I explained how to use *ETL* values in Section 6.7.

STUDY AIDS

With the material in this section, readers may measure their comprehension of this book. I recommend that you review the terms listed here for a chapter after reading that chapter. Then complete the self-examination for that chapter.

TERMS

Review the following terms. Rote memorization of definitions is less important than understanding the significance of the words.

Chapter 1

AISC
ASD
ASTM
Cold-forming
Hot-rolling

LRFD
Miscellaneous metals
Service load
Structural steel

Chapter 2

Allowable stress
Ductility
Modulus of elasticity
Plastic range
Rolled shapes
Stiffness
Strain hardening
Ultimate limit
Yield point

Chapter 3

Deformation limits
Field assemblage
Shop assemblage
Stability

Chapter 5

Combined load
Continuous beam
Cut section
Deformed shape
Factored load
Free-body diagram
Moment-resistive joint (connection)
Resistance factor
Rigid frame
Safety
Strength reduction factor
Structural investigation

Chapter 6

Bearing
Deck
Deflection
Elastic buckling
Equivalent uniform load (EUL)
Formed sheet steel deck
Framing: system, layout, plan
Inelastic buckling
Joist girder
Lateral buckling
Lateral unsupported length
Lightest section
Open-web steel joist
Plastic hinge
Safe load
Shear center
Steel joist (truss form)
Superimposed load
Torsional buckling
Web crippling
Web stiffener
Wide-flange beam

Chapter 7

Bending factor
Column interaction
Double-angle
Effective column buckling length
P-delta effect
Radius of gyration
Slenderness (L/r) ratio
Strut

Chapter 8

Bent
Captive column
Eccentric bracing
Rigid-frame bent
Sidesway
Trussed bent

Chapter 9

Composite structural element
Flitched beam
Net section

Chapter 10

Chord member
Maxwell diagram
Method of joints
Truss panel
Two-way spanning structure

Chapter 11

Boxing
Butt joint
Double shear
Edge distance
Effective area (in tension)
Effective length (of weld)
Fastener
Fillet weld
Framed beam connection
Gage (for angles)
Groove weld
Lap joint
Penetration (of weld)

Pitch
Plug weld
Single shear
Slot weld
Tearing
Tee joint (weld)
Throat (of weld)
Unfinished bolts (A307)
Upset end

Chapter 12

Building code
Dead load
Live load
Live load reduction

SELF-EXAMINATIONS

Questions

Note: Answers follow the last question.

Chapters 2–4

1. Why is steel's yield point more important than its ultimate strength for rolled structural products?

2. Why is the depth indicated in a W shape designation (e.g., 12 in. in W 12 × 36) known as a nominal dimension?

3. Steel is vulnerable in exposed conditions. What are a designer's primary concerns in this situation?

Chapter 5

1. What do the following illustrate?
   a. A free-body diagram
   b. A cut section

Chapter 6

1. What beam cross section property best predicts bending strength?

2. What is significant about L_c and L_u?

3. For A36 steel shapes used as beams, why do all beams with the same depth have the same deflection at their limiting loads on a given span?

4. What are the basic forms of buckling for a slender, unbraced beam?

5. What occurs when the loads are not in the plane of the beam cross section's shear center?

6. What is the most common way to prevent torsion on a W shape beam?

7. What cross section property is most critical to resisting web crippling in a W shape beam?

Chapter 7

1. When evaluating simple axial compression capacity, what are the significant properties of a steel column cross section?

2. When must you consider how buckling affects both axes of a W shape steel column?

3. How do end support conditions affect column buckling?

4. What is the P-delta effect?

Chapter 8

1. What interaction is required between rigid-frame members?

2. When structural frames and structural walls interact, why do the walls tend to take most of the load?

3. Why is wind load not always a critical concern for design of individual structural members?

4. Regarding deformation of a rigid frame, what is significant about the relative stiffness of the frame members?

Chapter 10

1. Why should trusses be loaded only at their panel points?

Chapter 11

1. When you use high-strength steel bolts to connect steel members, what basic action develops the initial load resistance in the joint?

2. What stresses combine to resist tearing in a bolted connection?

3. Other than spacing and edge distances, what basic dimension limits the number of bolts you can use in a framed beam connection?

4. Why shouldn't supporting steel beams have the same depth as the beams they support?

5. Why is the throat dimension in a fillet weld significant?

6. Boxing welds in a joint gains what structural advantage?

Chapter 12

1. What is the design dead load?

2. What factor most affects live load reduction?

Answers

Chapters 2–4

1. A structure's practical limit is usually its deformation in ductile response, so acceptable performance is based on yield.

2. In most cases, a shape's true depth is not the designated depth.

3. Rusting; failure in fire.

Chapter 5

1. **a.** The loads and reactions; that is, the general exterior forces required for stability.
   b. The internal force actions, stresses, and local deformations.

Chapter 6

1. Section modulus.

2. They indicate limits with respect to lateral unsupported conditions.

3. Strain is proportional to stress, which is proportional to load magnitude. For the same depth, deflection is proportional to stress. Thus all beams with the same depth have the same deflection if stress (maximum allowable) is a constant.

4. Lateral (sideways) buckling of the compression side of the beam, in column action. Rotational (torsional) buckling at midspan or at supports.

5. Twisting; a torsional moment on the beam.

6. Brace the beam.

7. Beam web thickness.

Chapter 7

1. Area; radius of gyration.

2. When effective buckling length (KL) is different on the two axes.

3. They may alter the effective length of the column that forms the buckled profile, thus changing the column's resistance to buckling.

4. The P-delta effect is a bending moment (P times column deformation) that produces more deflection (i.e., additional delta), which results in greater P-delta, and so on.

Chapter 8

1. Transfer of bending moment (through connections).

2. They are typically much stiffer than the frames in resisting lateral forces.

3. The lowered safety factor used with wind loads may reduce the design load for the combined wind and gravity loading below the design load for gravity loading only. Because of this, it is not a critical design concern.

4. If some members are relatively exceptionally stiff or flexible, it may cause a different form of general deformation of the frame.

Chapter 10

1. To avoid shear and bending in truss chord members.

Chapter 11

1. Friction between the connected parts, induced by the clamping (squeezing) action of the highly tensioned bolts.

2. Shear and tension.

3. Beam depth.

4. Because you must cut back both flanges of the supported beam to achieve the connection (see Figure 11.13*f*). Doing so results in significant loss of end shear capacity.

5. It defines the critical weld cross section for shear stress, which in turn defines the weld strength.

6. Increased resistance to tearing caused by twisting actions on the welded joint.

Chapter 12

1. The weight of the building construction and the weights of permanently attached items.

2. The roof and floor area supported by a structural member.

ANSWERS TO PROBLEMS

CHAPTER 5

5.4.A. R = 10 kips up and 110 kip-ft counterclockwise
5.4.B. R = 5 kips up and 24 kip-ft counterclockwise
5.4.C. R = 6 kips to the left and 72 kip-ft counterclockwise
5.4.D. Left R = 4.5 kips up, right R = 4.5 kips down and 12 kips to the right
5.4.E. Left R = 4.5 kips down and 6 kips to the left, right R = 4.5 kips up and 6 kips to the left

CHAPTER 6

6.2.A. W 10 × 19 or W 12 × 19
6.2.B. W 12 × 22 or W 14 × 22
6.2.C. W 10 × 19 or W 12 × 19
6.2.D. W 24 × 55

6.2.E. W 12 × 26 (U.S.); W 14 × 22 (metric)

6.2.F. W 16 × 31

6.2.G. W 10 × 19 or W 12 × 19

6.2.H. W 12 × 22 or W 14 × 22

6.2.I. W 14 × 22

6.2.J. W 16 × 26

6.2.K. 13.6%

6.2.L. 51.5%

6.4.A. 0.80 in.

6.4.B. 0.69 in.

6.4.C. 0.83 in.

6.5.A. a) W 30 × 90 b) W 30 × 99 c) W 21 × 111

6.5.B. a) W 21 × 68 b) W 24 × 68 c) W 24 × 76

6.9.A. 42.4 kips

6.9.B. Probably, end bearing computation yields only 48.5 kips

6.10.A. B = 15 in., t = 1 in.

6.11.A. 24K9

6.11.B. 26K9

6.11.C. a) 24K4 b) 20K7

6.11.D. a) 20K3 b) 16K6

6.12.A. WR20

6.12.B. WR18

6.12.C. IR22 or WR22

CHAPTER 7

7.5.A. 430 kips [1912 kN]

7.5.B. 278 kips [1237 kN]

7.5.C. 375 kips [1669 kN]

7.7.A. W 8 × 31

7.7.B. W 12 × 58

7.7.C. W 12 × 79

7.8.A.   4 in.

7.8.B.   5 in.

7.8.C.   6 in.

7.8.D.   8 in.

7.10.A.   78 kips

7.10.B. $4 \times 3 \times \frac{5}{16}$ or $3\frac{1}{2} \times 2\frac{1}{2} \times \frac{3}{8}$

7.11.A.   W 12×58

7.11.B.   W 14×120

7.13.A.   15 in. $\times$ 16 in. $\times$ $1\frac{1}{4}$ in. (required $t = 1.235$ in.)

7.13.B.   10 in. $\times$ 12 in. $\times$ 1 in. (required $t = 0.89$ in.)

CHAPTER 9

9.2.A.   21.2 kips

9.3.A.   Reaction: $V = 5$ kips, $H = 12.5$ kips; tension in cable is 13.46 kips

9.3.B.   At the left support: $V = 6.67$ kips, $H = 6.67$ kips; at the right reaction: $V = 3.33$ kips, $H = 6.67$ kips; maximum tension in cable = 9.43 kips

CHAPTER 10

10.4.A.   Sample values: $CI = 2000C$, $IJ = 812.5T$, $JG = 1250T$

10.4.B.   Sample values: $CJ = 2828C$, $JK = 1118T$, $KH = 1500T$

10.4.C.   Same as Problem 10.4.A

10.4.D.   Same as Problem 10.4.B

CHAPTER 11

11.4.A.   Outer plates: $\frac{5}{8}$ in., middle plate: $\frac{3}{4}$ in.

11.4.B.   Same as Problem 11.4.A

11.6.A.   Rounding off, use $L_1 = 11$ in., $L_2 = 5$ in.

11.6.B.   Minimum of 4.5-in.-long welds on each side

BIBLIOGRAPHY

1. *Uniform Building Code*, Vol. 2, *Structural Engineering Provisions*, 1994 ed., International Conference of Building Officials (ICBO), Whittier, CA, 1994. (Known as the UBC.)

2. *Minimum Design Loads for Buildings and Other Structures* ANSI/ASCE 7-88, American Society of Civil Engineers (ASCE), New York, 1990 (revision of ANSI A58.1-1982, a publication by American National Standards Institute (ANSI).

3. *Manual of Steel Construction: Allowable Stress Design*, 9th ed., American Institute of Steel Construction (AISC), Chicago, 1989. (Known as the AISC Manual.)

4. *Manual of Steel Construction: Load and Resistance Factor Design*, American Institute of Steel Construction (AISC), Chicago, 1986.

5. *Steel Buildings: Analysis and Design*, 4th ed., Stanley W. Crawley and Robert M. Dillon, John Wiley and Sons, New York, 1993.

6. *Structural Steel Design: LRFD Method*, Jack C. McCormac, Harper Collins, New York, 1995.

7. *Standard Specifications, Load Tables, and Weight Tables for Steel Joists and Joist Girders*, Steel Joist Institute (SJI), Myrtle Beach, SC, 1989.

8. *Steel Deck Institute Design Manual for Composite Decks, Form Decks, and Roof Decks*, Steel Deck Institute, St. Louis, MO, 1982.

9. *Simplified Building Design for Wind and Earthquake Forces*, 3rd ed., James Ambrose and Dimitry Vergun, John Wiley and Sons, New York, 1995.

10. *Architectural Graphic Standards*, 9th ed., Charles G. Ramsey and Harold R. Sleeper, John Wiley and Sons, New York, 1994.

11. *Fundamentals of Building Construction: Materials and Methods*, 2nd ed., Edward Allen, John Wiley and Sons, New York, 1990.

12. *Standard Handbook of Structural Details for Building Construction*, 2nd ed., Morton Newman, McGraw-Hill, New York, 1993.

13. *Construction Revisited*, James Ambrose, John Wiley & Sons, New York, 1993.

14. *Design of Building Trusses*, James Ambrose, John Wiley and Sons, New York, 1994.

INDEX